Linear Algebra

Linear Algebra

Juan Jorge Schäffer

Carnegie Mellon University, USA

W⦿ World Scientific

NEW JERSEY · LONDON · SINGAPORE · BEIJING · SHANGHAI · HONG KONG · TAIPEI · CHENNAI

Published by

World Scientific Publishing Co. Pte. Ltd.

5 Toh Tuck Link, Singapore 596224

USA office: 27 Warren Street, Suite 401-402, Hackensack, NJ 07601

UK office: 57 Shelton Street, Covent Garden, London WC2H 9HE

Library of Congress Cataloging-in-Publication Data
Schäffer, Juan Jorge.
    Linear algebra / by Juan Jorge Schäffer (Carnegie Mellon University, USA).
    pages cm
    Includes indexes.
    ISBN 978-9814623490 (hardcover : alk. paper)
    1. Algebras, Linear. 2. Algebra. I. Title.
    QA184.2.S31 2014
    512'.5--dc23
                                        2014026825

British Library Cataloguing-in-Publication Data
A catalogue record for this book is available from the British Library.

LATEX composition and PICTEX figures: Nancy J. Watson

Printed in Singapore

PREFACE

This book is a companion volume to the author's *Basic Language of Mathematics* and is inspired by the same concern to provide a clear, comprehensive, and formally sound presentation of its subject matter. The work is also intended to provide a reliable bridge to broader areas of mathematics, such as are, for instance, addressed by the author's long-standing mentor and collaborator Walter Noll in his *Finite-Dimensional Spaces, Algebra, Geometry, and Analysis Volume I*, a book developed in the same academic environment.

Notation, terminology, and basic results on sets, mappings, families, relations, ordered sets, and natural and real numbers, as well as some elementary facts about commutative monoids, groups, and rings (including fields) are taken directly from *Basic Language of Mathematics*; explicit references to that work use the abbreviation *Basic Language*.

The author thanks Ms Nancy J. Watson for her excellent work on the preparation of the manuscript.

CONTENTS

Chapter 1

LINEAR SPACES AND LINEAR MAPPINGS

11. Linear spaces

Throughout this work we shall assume that a field is given. In order to make plain that this field is fixed, we denote it by $\mathbb{F}$. The reader may wish to imagine that $\mathbb{F} := \mathbb{R}$; this may suggest a geometric view of some matters. It is, however, with few exceptions (to be noted), quite immaterial what particular field $\mathbb{F}$ is. It is also customary to denote the unity of the field $\mathbb{F}$ by 1. If $\mathbb{F} := \mathbb{R}$, this is of course right; in general, we shall see that it will not lead to a notational clash, any more than using 0 for the zero of $\mathbb{F}$ will. We must, however, realize that it may then happen, e.g., that $1 + 1 = 0$ (cf. *Basic Language*, Example 133A,(c)).

We define a **linear space** (**over** $\mathbb{F}$) to be a commutative group $\mathcal{V}$, written additively, endowed with additional structure by the prescription of a family of mappings $((u \mapsto su) \mid s \in \mathbb{F}) \in (\mathrm{Map}(\mathcal{V}, \mathcal{V}))^{\mathbb{F}}$, called the **scalar multiplication**, subject to the following conditions:

(LS1): $\forall s \in \mathbb{F}, \ \forall u, v \in \mathcal{V}, \quad s(u + v) = (su) + (sv) \quad$ *(space distributive law)*

(LS2): $\forall s, t \in \mathbb{F}, \ \forall u \in \mathcal{V}, \quad (s + t)u = (su) + (tu) \quad$ *(field distributive law)*

(LS3): $\forall s, t, \in \mathbb{F}, \ \forall u \in \mathcal{V}, \quad (st)u = s(tu) \quad$ *(composition law)*

(LS4): $\forall u \in \mathcal{V}, \quad 1u = u \quad$ *where 1 is the unity of* $\mathbb{F}$ *(unity law)*.

One often finds the term *vector space* used instead of *linear space*. The members of $\mathbb{F}$ are sometimes referred to as *scalars* and the members of a linear space $\mathcal{V}$ as *vectors*; it is perhaps best to reserve that terminology for a specialized context. If $s \in \mathbb{F}$ and $u \in \mathcal{V}$, we say that su is u **multiplied by** s, and read 's **times** u''; we call su a **scalar multiple of** u.

The usual conventions about the use of parentheses are adopted; thus, scalar multiplication has priority over addition, opposition, and subtraction, so that the right-hand sides of (LS1), (LS2) may be written without parentheses. By virtue of (LS3) we may also write stu, without parentheses, for $s, t \in \mathbb{F}$, $u \in \mathcal{V}$.

To the notations used for commutative monoids and groups we add the following. If K is a subset of $\mathbb{F}$ and $\mathcal{A}$ is a subset of $\mathcal{V}$, we set

$$K\mathcal{A} := \{su \mid (s, u) \in K \times \mathcal{A}\} \subset \mathcal{V};$$

and if $t \in \mathbb{F}$, $v \in \mathcal{V}$, we set $t\mathcal{A} := \{t\}\mathcal{A}$ and $Kv := K\{v\}$.

It is sometimes necessary to have names for the zero, addition, opposition, and scalar multiplication in a specific linear space $\mathcal{V}$. In that case we use $0^{\mathcal{V}}$, $\mathrm{add}^{\mathcal{V}}$, $\mathrm{opp}^{\mathcal{V}}$, $\mathrm{mult}^{\mathcal{V}}$, respectively. Thus, e.g.,

$$\mathrm{mult}^{\mathcal{V}}_s(u) := su \quad \text{for all } s \in \mathbb{F} \text{ and } u \in \mathcal{V}.$$

With these notations, (LS1)–(LS4) become

(LS1): $\forall s \in \mathbb{F}, \quad \mathrm{mult}^{\mathcal{V}}_s \circ \mathrm{add}^{\mathcal{V}} = \mathrm{add}^{\mathcal{V}} \circ (\mathrm{mult}^{\mathcal{V}}_s \times \mathrm{mult}^{\mathcal{V}}_s)$

(LS2): $\forall s, t \in \mathbb{F}, \quad \mathrm{mult}^{\mathcal{V}}_{s+t} = \mathrm{add}^{\mathcal{V}} \circ (\mathrm{mult}^{\mathcal{V}}_s, \mathrm{mult}^{\mathcal{V}}_t)$

(LS3): $\forall s, t \in \mathbb{F}, \quad \mathrm{mult}^{\mathcal{V}}_{st} = \mathrm{mult}^{\mathcal{V}}_s \circ \mathrm{mult}^{\mathcal{V}}_t$

(LS4): $\mathrm{mult}^{\mathcal{V}}_1 = 1_{\mathcal{V}}$.

We record some elementary facts about the operations in a linear space.

11A. PROPOSITION. *Let the linear space $\mathcal{V}$ be given. Then*

(11.1) $$\forall s \in \mathbb{F}, \quad s0 = 0 \quad \text{where } 0 := 0^{\mathcal{V}}$$

(11.2) $$\forall u \in \mathcal{V} \quad 0u = 0^{\mathcal{V}} \quad \text{where } 0 \text{ is the zero of } \mathbb{F}$$

(11.3) $$\forall s \in \mathbb{F}, \quad \forall u \in \mathcal{V}, \quad (-s)u = s(-u) = -su \text{ and } (-s)(-u) = su$$

(11.4) $$\forall s \in \mathbb{F}, \quad \forall u, v \in \mathcal{V}, \quad s(u - v) = su - sv$$

(11.5) $$\forall s, t \in \mathbb{F}, \quad \forall u \in \mathcal{V}, \quad (s - t)u = su - tu$$

(11.6) $$\forall u \in \mathcal{V}, \quad (-1)u = -u.$$

Formulas (11.2) and (11.6) say

(11.7) $$\mathrm{mult}^{\mathcal{V}}_0 = (0^{\mathcal{V}})_{\mathcal{V} \to \mathcal{V}} \qquad \mathrm{mult}^{\mathcal{V}}_{-1} = \mathrm{opp}^{\mathcal{V}}.$$

Proof. The proof of (11.1)–(11.5) is quite similar in form to the proof of *Basic Language*, Propositions 132B, and we omit it. By (11.5), (11.2), (LS4) we have

$$(-1)u = (0 - 1)u = 0u - 1u = 0 - u = -u$$

for all $u \in \mathcal{V}$, so that (11.6) holds. ■

On comparing (11.2) and (LS4) with *Basic Language*, (115.2), we notice that there is no clash between $0u$ and $1u$, meaning u multiplied by the zero and the unity of $\mathbb{F}$, on the one hand, and $0u$ and $1u$, meaning the 0th and the 1st multiple of u, on the other.

We also require generalized versions of the distributive laws.

11B. PROPOSITION. *Let the linear space $\mathcal{V}$ be given.*

(a): *Let the family $z \in \mathcal{V}^I$ and $s \in \mathbb{F}$ be given. Then* $\mathrm{Supp}\{sz_i \mid i \in I\} \subset \mathrm{Supp}z$. *For every finite subset J of I, and for every subset J of I if $\mathrm{Supp}z$ is finite, we have*

$$s \sum_J z = \sum_{j \in J} sz_j.$$

(b): *Let $u \in \mathcal{V}$ and the family $a \in \mathbb{F}^K$ be given. Then* $\mathrm{Supp}\{a_k u \mid k \in K\} \subset$ $\subset \mathrm{Supp}a$. *For every finite subset L of K, and for every subset L of K if $\mathrm{Supp}a$ is finite, we have*

$$\left(\sum_L a\right)u = \sum_{l \in L} a_l u.$$

(c): *Let the families $z \in \mathcal{V}^I$ and $a \in \mathbb{F}^K$ be given. Then* $\mathrm{Supp}\{a_k z_i \mid (k,i) \in K \times I\} \subset \mathrm{Supp}a \times \mathrm{Supp}z$. *For all finite subsets J of I and L of K, and for all subsets J of I and L of K if $\mathrm{Supp}z$ and $\mathrm{Supp}a$ are finite, we have*

$$\left(\sum_L a\right)\left(\sum_J z\right) = \sum_{(l,j) \in L \times J} a_l z_j.$$

(d): $\forall s \in \mathbb{F}, \; \forall u \in \mathcal{V}, \; \forall m, n \in \mathbb{N}, \quad (ms)(nu) = (mn)(su);$ *in particular,*

(11.8) $\qquad \forall u \in \mathcal{V}, \; \forall m \in \mathbb{N}, \quad (m1)u = mu, \quad$ *where 1 is the unity of $\mathbb{F}$.*

Proof. The proof is quite similar in form to the proof of *Basic Language*, Proposition 132D, and we therefore omit it. ■

According to Proposition 11B,(d), we may omit parentheses in expressions such as msu for $m \in \mathbb{N}$, $s \in \mathbb{F}$, $u \in \mathcal{V}$. It follows from (11.8) that if, for each $m \in \mathbb{N}$, the mth multiple of the unity of $\mathbb{F}$ happens to be denoted by m, there is no clash between mu, meaning u multiplied by $m \in \mathbb{F}$, and mu, meaning the mth multiple of u. This is applicable, in particular, when $\mathbb{F} := \mathbb{R}$ or $\mathbb{F} := \mathbb{Q}$, where the mth multiple of the unity of $\mathbb{F}$ actually is the natural number m itself.

We next establish a converse to the conjunction of (11.1) and (11.2).

11C. PROPOSITION. *Let the linear space $\mathcal{V}$ be given. Then* $\mathbb{F}^\times \mathcal{V}^\times \subset \mathcal{V}^\times$, *i.e.,*

$$\forall s \in \mathbb{F}, \; \forall u \in \mathcal{V}, \quad su = 0 \Rightarrow (s = 0 \; \text{or} \; u = 0).$$

Proof. Let $s \in \mathbb{F}^\times$ and $u \in \mathcal{V}^\times$ be given. By (LS3), (LS4) we have $\frac{1}{s}(su) = (\frac{1}{s}s)u = 1u = u \in \mathcal{V}^\times$. It follows from (11.1) that $su \neq 0$, i.e., $su \in \mathcal{V}^\times$. ■

11D. EXAMPLES. (a): Every singleton can be endowed with the structure of a linear space over $\mathbb{F}$ (in exactly one way) by the one unavoidable choice of zero, addition, opposition, and scalar multiplication. Such a space has then the form $\{0\}$, and is called a **zero-space**. A linear space is said to be **trivial** if it is a zero-space.

(b): $\mathbb{F}$ itself, regarded as its additive group, becomes a linear space over the field $\mathbb{F}$ with the scalar multiplication defined by the rule

$$\text{mult}^{\mathbb{F}}{}_s(t) := st \quad \text{for all } s, t \in \mathbb{F},$$

where st is the product of s and t in the field $\mathbb{F}$. The validity of (LS1) and (LS2) follows from the commutative law for multiplication and the distributive law in the field $\mathbb{F}$; the validity of (LS3) and (LS4) follows, respectively, from the associative law and the neutrality law for multiplication in $\mathbb{F}$. When we refer to $\mathbb{F}$ as a linear space over the field $\mathbb{F}$, we shall always mean this particular linear-space structure.

(c): Let the family $(\mathcal{V}_i \mid i \in I)$ of linear spaces be given. Then the set $\bigtimes_{i\in I} \mathcal{V}_i$, endowed with structure by the prescription of zero, addition, opposition, and scalar multiplication defined *termwise*, i.e., by the rules

$$0 := (0 \mid i \in I)$$

$$\forall a, b \in \bigtimes_{i\in I} \mathcal{V}_i, \quad a + b := (a_i + b_i \mid i \in I)$$

$$\forall a \in \bigtimes_{i\in I} \mathcal{V}_i, \quad -a := (-a_i \mid i \in I)$$

$$\forall s \in \mathbb{F}, \ \forall a \in \bigtimes_{i\in I} \mathcal{V}_i, \quad sa := (sa_i \mid i \in I),$$

is a linear space. We shall always regard $\bigtimes_{i\in I} \mathcal{V}_i$ as endowed with this linear-space structure. It is called the (**Cartesian**) **product of** the family $(\mathcal{V}_i \mid i \in I)$. For every $a \in \bigtimes_{i\in I} \mathcal{V}_i$ we define the set $\text{Supp}\, a := \{i \in I \mid a_i \neq 0\}$, called the **support of** a. It is also useful to define the set

$$\text{Supp}(\mathcal{V}_i \mid i \in I) := \{i \in I \mid \mathcal{V}_i \neq \{0\}\} = \bigcup \{\text{Supp}\, a \mid a \in \bigtimes_{i\in I} \mathcal{V}_i\},$$

and call it the **support of** the family $(\mathcal{V}_i \mid i \in I)$.

In particular, if I is a set and $\mathcal{V}$ is a linear space, we have the Cartesian product $\mathcal{V}^I := \bigtimes_{i\in I} \mathcal{V}$ of the family $(\mathcal{V} \mid i \in I)$. For this family, the definitions of zero, addition, and support agree with the definitions for monoids of families given in *Basic Language*, Section 117. We note that $\mathcal{V}^I$ is a zero-space if and only if I is empty or $\mathcal{V}$ is a zero-space. Using (b), we have in particular the linear spaces $\mathbb{F}^I$ for all sets I.

As a kind of special case we also have the following. Let the linear spaces $\mathcal{V}$ and $\mathcal{W}$ be given. Then the operations on the **product** $\mathcal{V} \times \mathcal{W}$ **of** $\mathcal{V}$ and $\mathcal{W}$ are given by the rules

$$0 := (0,0)$$
$$\forall (v,w),(v',w') \in \mathcal{V} \times \mathcal{W}, \quad (v,w)+(v',w') := (v+v',w+w')$$
$$\forall (v,w) \in \mathcal{V} \times \mathcal{W}, \quad -(v,w) := (-v,-w)$$
$$\forall s \in \mathbb{F}, \forall (v,w) \in \mathcal{V} \times \mathcal{W}, \quad s(v,w) := (sv,sw).$$

(d): Let the set D and the linear space $\mathcal{V}$ be given. Then the set $\mathrm{Map}(D,\mathcal{V})$, endowed with structure by the prescription of zero, addition, opposition, and scalar multiplication defined *valuewise*, i.e., by the rules

$$0 := 0$$
$$\forall f,g \in \mathrm{Map}(D,\mathcal{V}), \ \forall x \in D, \quad (f+g)(x) := f(x)+g(x)$$
$$\forall f \in \mathrm{Map}(D,\mathcal{V}), \ \forall x \in D, \quad (-f)(x) := -f(x)$$
$$\forall s \in \mathbb{F}, \ \forall f \in \mathrm{Map}(D,\mathcal{V}), \ \forall x \in D, \quad (sf)(x) := sf(x),$$

is a linear space. We shall always regard $\mathrm{Map}(D,\mathcal{V})$ as endowed with this linear-space structure. Actually, this space is a variant of the space $\mathcal{V}^D$ defined according to (c).
∎

12. Linear mappings

Let the linear spaces $\mathcal{V}$, $\mathcal{W}$ be given. A mapping $L \colon \mathcal{V} \to \mathcal{W}$ is said to be **linear** if it satisfies the following conditions:

$$(12.1) \qquad\qquad L \circ \mathrm{add}^{\mathcal{V}} = \mathrm{add}^{\mathcal{W}} \circ (L \times L)$$

$$(12.2) \qquad\qquad L(0^{\mathcal{V}}) = 0^{\mathcal{W}}$$

$$(12.3) \qquad\qquad L \circ \mathrm{opp}^{\mathcal{V}} = \mathrm{opp}^{\mathcal{W}} \circ L$$

$$(12.4) \qquad\qquad \forall s \in \mathbb{F}, \quad L \circ \mathrm{mult}^{\mathcal{V}}_{s} = \mathrm{mult}^{\mathcal{W}}_{s} \circ L.$$

Roughly speaking, a mapping is linear if it preserves the linear-space structure.

12A. Remark. (a): Using (11.7), we see that (12.2) and (12.3) are nothing but the special cases of (12.4) for $s := 0$ and $s := -1$, respectively, and are therefore redundant. In order to verify that a mapping $L : \mathcal{V} \to \mathcal{W}$ is linear, it is therefore sufficient to establish that L satisfies (12.1) and (12.4), i.e.,

$$(12.5) \qquad\qquad \forall u, v \in \mathcal{V}, \; L(u + v) = L(u) + L(v)$$

$$(12.6) \qquad\qquad \forall s \in \mathbb{F}, \quad \forall u \in \mathcal{V}, \quad L(su) = sL(u).$$

A mapping that satisfies (12.1) or, equivalently, (12.5) is said to be **additive**, and one that satisfies (12.4) or, equivalently, (12.6) is said to be **homogeneous**. These conditions are expressed, respectively, by the commutativity of the following diagrams (the second one for each $s \in \mathbb{F}$):

(b): Let the linear space $\mathcal{V}$ be given. Then $\mathrm{mult}^{\mathcal{V}}_{t} : \mathcal{V} \to \mathcal{V}$ is a linear mapping for every $t \in \mathbb{F}$: the additivity follows from (LS1), the homogeneity from (LS3) and the commutative law for multiplication in $\mathbb{F}$. In particular, $1_{\mathcal{V}} = \text{mult}^{\mathcal{V}}{}_1$ and $\text{opp}^{\mathcal{V}} = \text{mult}^{\mathcal{V}}{}_{-1}$ are linear.

(c): For given linear spaces $\mathcal{V}$, $\mathcal{W}$, (12.2) shows that the only constant mapping from $\mathcal{V}$ to $\mathcal{W}$ that is linear is $0_{\mathcal{V} \to \mathcal{W}}$, the **zero-mapping**. ∎

12B. PROPOSITION. *Let the linear mapping $L : \mathcal{V} \to \mathcal{W}$ and the family $z \in \mathcal{V}^I$ be given. Then $\text{Supp}(L \circ z) \subset \text{Supp} z$. For every finite subset J of I, and for every subset J of I if $\text{Supp} z$ is finite, we have*

$$L\left(\sum_J z\right) = \sum_J L \circ z.$$

Proof. The proof is again quite similar in form to the proof of *Basic Language,* Proposition 132D, and we therefore omit it. ∎

For given linear spaces $\mathcal{V}$, $\mathcal{W}$, we set

$$\text{Lin}(\mathcal{V}, \mathcal{W}) := \{L \in \text{Map}(\mathcal{V}, \mathcal{W}) \mid L \text{ is linear}\}.$$

In particular, for every linear space $\mathcal{V}$ we set $\text{Lin}\mathcal{V} := \text{Lin}(\mathcal{V}, \mathcal{V})$

12C. PROPOSITION. *Let the linear spaces $\mathcal{V}$, $\mathcal{W}$, $\mathcal{X}$, and the linear mappings $L : \mathcal{V} \to \mathcal{W}$ and $M : \mathcal{W} \to \mathcal{X}$ be given. Then $M \circ L$ is linear.*

Proof. Since L and M are linear, we have

$$M \circ L \circ \text{add}^{\mathcal{V}} = M \circ \text{add}^{\mathcal{W}} \circ (L \times L) = \text{add}^{\mathcal{X}} \circ (M \times M) \circ (L \times L) =$$
$$= \text{add}^{\mathcal{X}} \circ ((M \circ L) \times (M \circ L))$$

$$M \circ L \circ \text{mult}^{\mathcal{V}}{}_s = M \circ \text{mult}^{\mathcal{W}}{}_s \circ L = \text{mult}^{\mathcal{X}}{}_s \circ M \circ L \quad \text{for all } s \in \mathbb{F},$$

as was to be shown. ∎

The computations in the preceding proof are expressed by "diagram-chasing" as follows:

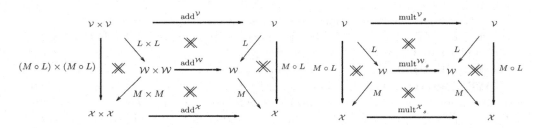

12D. PROPOSITION. *Let the linear spaces $\mathcal{V}$, $\mathcal{W}$, $\mathcal{X}$, and the mappings $L : \mathcal{V} \to \mathcal{W}$ and $M : \mathcal{W} \to \mathcal{X}$ be given.*

(a): *If L and $M \circ L$ are linear and L is surjective, then M is linear.*

(b): *If M and $M \circ L$ are linear and M is injective, then L is linear.*

Proof. Proof of (a). Since L and $M \circ L$ are linear, we have

$$M \circ \mathrm{add}^{\mathcal{W}} \circ (L \times L) = M \circ L \circ \mathrm{add}^{\mathcal{V}} = \mathrm{add}^{\mathcal{X}} \circ ((M \circ L) \times (M \circ L)) =$$
$$= \mathrm{add}^{\mathcal{X}} \circ (M \times M) \circ (L \times L)$$
$$M \circ \mathrm{mult}^{\mathcal{W}}{}_s \circ L = M \circ L \circ \mathrm{mult}^{\mathcal{V}}{}_s = \mathrm{mult}^{\mathcal{X}}{}_s \circ M \circ L \quad \text{for all } s \in \mathbb{F}.$$

Since L is surjective, so is $L \times L : \mathcal{V} \times \mathcal{V} \to \mathcal{W} \times \mathcal{W}$, and hence both L and $L \times L$ are right-cancelable (*Basic Language*, Proposition 35A.R). It follows that

$$M \circ \mathrm{add}^{\mathcal{W}} = \mathrm{add}^{\mathcal{X}} \circ (M \times M)$$

$$M \circ \mathrm{mult}^{\mathcal{W}}{}_s = \mathrm{mult}^{\mathcal{X}}{}_s \circ M \quad \text{for all } s \in \mathbb{F},$$

so that M is linear.

Proof of (b). Since M and $M \circ L$ are linear, we have

$$M \circ L \circ \mathrm{add}^{\mathcal{V}}$$
$$= \mathrm{add}^{\mathcal{X}} \circ ((M \circ L) \times (M \circ L)) = \mathrm{add}^{\mathcal{X}} \circ (M \times M) \circ (L \times L) =$$
$$= M \circ \mathrm{add}^{\mathcal{W}} \circ (L \times L)$$
$$M \circ L \circ \mathrm{mult}^{\mathcal{V}}{}_s = \mathrm{mult}^{\mathcal{X}}{}_s \circ M \circ L = M \circ \mathrm{mult}^{\mathcal{W}}{}_s \circ L \quad \text{for all } s \in \mathbb{F}.$$

Since M is injective, M is left-cancelable (*Basic Language*, Proposition 35A.L). It follows that L satisfies (12.1) and (12.4), and is therefore linear. ∎

12E. COROLLARY. *If a linear mapping is bijective, its inverse is linear.*

Proof. Let the bijective linear mapping L be given. Then L is linear and injective, and $L \circ L^{\leftarrow} = 1_{\mathrm{Cod}L}$ is linear (Remark 12A, (b)). By Proposition 12D,(b), $L^{\leftarrow}$ is linear. ∎

There are some notational conventions in common use for linear mappings. Let the linear mapping L be given. We write $Lu := L(u)$ for all $u \in \mathrm{Dom}L$, $Lf := L \circ f$ for all mappings f with $\mathrm{Cod}f = \mathrm{Dom}L$, and $Lz := L \circ z$ for all families z with $\mathrm{Rng}z \subset \mathrm{Dom}L$. These conventions yield such unambiguous expressions as MLu, where M and L are linear mappings with $\mathrm{Cod}L = \mathrm{Dom}M$, and $u \in \mathrm{Dom}L$. (Note that ML is linear, by Proposition 12C.) If L is a linear mapping with $\mathrm{Dom}L = \mathrm{Cod}L$, we write $L^n := L^{\circ n}$ for all $n \in \mathbb{N}$. If L is a bijective linear mapping, so that its inverse is also linear by Corollary 2E, we write $L^{-1} := L^{\leftarrow}$. Finally, if L is a bijective linear mapping with $\mathrm{Dom}L = \mathrm{Cod}L$, we write, more generally, $L^{-n} := (L^{\leftarrow})^n$ for all $n \in \mathbb{N}^{\times}$.

12F. REMARKS. Let the linear mapping L with $\mathrm{Dom}L = \mathrm{Cod}L$ be given. It follows from *Basic Language*, Propositions 96B and 97C that $L^{m+n} = L^m L^n$ and $L^{mn} = (L^m)^n$ for all $m, n \in \mathbb{N}$. Suppose now that L is also bijective. It follows from *Basic Language*, Proposition 94D that $L^{-n} = (L^{\leftarrow})^n = (L^n)^{\leftarrow}$ for all $n \in \mathbb{N}$. Combining these facts with some computation, it then follows that $L^{m+n} = L^m L^n$ and $L^{mn} = (L^m)^n$ for all $m, n \in \mathbb{Z}$; we thus see that the notation $L^{-n} := (L^{\leftarrow})^n$ for all $n \in \mathbb{N}^{\times}$ does not lead to a clash. ∎

12G. EXAMPLES. (a): Let the family of linear spaces $(\mathcal{V}_i \mid i \in I)$ be given, and consider the Cartesian product $\bigtimes_{i \in I} \mathcal{V}_i$ as defined in Example 11D,(c). For each $j \in I$

the projection $\pi_j : \bigtimes_{i \in I} \mathcal{V}_i \to \mathcal{V}_j$ is then a linear mapping, as follows at once from the definitions.

Similarly (or as a special case), for given linear spaces $\mathcal{V}$, $\mathcal{W}$ the mappings $((v, w) \mapsto v) : \mathcal{V} \times \mathcal{W} \to \mathcal{V}$ and $((v, w) \mapsto w) : \mathcal{V} \times \mathcal{W} \to \mathcal{W}$ are linear.

(b): Let the set D and the linear space $\mathcal{V}$ be given, and consider the linear space $\mathrm{Map}(D, \mathcal{V})$ defined in Example 11D,(d). For each $x \in D$ the evaluation mapping $\mathrm{ev}_x : \mathrm{Map}(D, \mathcal{V}) \to \mathcal{V}$, i.e., the mapping $(f \mapsto f(x)) : \mathrm{Map}(D, \mathcal{V}) \to \mathcal{V}$, is linear. It follows from this and from Proposition 12B that for every $x \in D$, every family $(f_i \mid i \in I)$ in $\mathrm{Map}(D, \mathcal{V})$, and every finite subset J of I – indeed, every subset J of I if $\mathrm{Supp}(f_i \mid i \in I)$ is finite, we have

$$\Big(\sum_{j \in J} f_j\Big)(x) = \sum_{j \in J} f_j(x).$$

(c): Let the family of linear spaces $(\mathcal{V}_k \mid k \in K)$, the set I, and the mapping $\phi : I \to K$ be given. Then the Mapping

$$a \mapsto a \circ \phi : \quad \bigtimes_{k \in K} \mathcal{V}_k \to \bigtimes_{i \in I} \mathcal{V}_{\phi(i)}$$

is linear.

(d): Let the sets D and C, the linear space $\mathcal{V}$, and the mapping $\phi : D \to C$ be given. Then the mapping

$$f \mapsto f \circ \phi : \mathrm{Map}(C, \mathcal{V}) \to \mathrm{Map}(D, \mathcal{V})$$

is linear.

(e): Let the non-empty set D, the linear spaces $\mathcal{V}$ and $\mathcal{W}$, and the mapping $L : \mathcal{V} \to \mathcal{W}$ be given. Then the mapping

$$f \mapsto L \circ f : \mathrm{Map}(D, \mathcal{V}) \to \mathrm{Map}(D, \mathcal{W})$$

is linear if and only if L is linear.

(f): Let the set I and the linear space $\mathcal{V}$ be given. For every finite subset J of I the mapping

$$a \mapsto \sum_J a : \mathcal{V}^I \to \mathcal{V}$$

is linear. This follows from *Basic Language*, Theorem 117E for the additivity, and from Proposition 11B,(a) for the homogeneity.

Similarly (or as a special case), the mapping $\mathrm{add}^{\mathcal{V}} : \mathcal{V} \times \mathcal{V} \to \mathcal{V}$ is linear for every linear space $\mathcal{V}$.

(g): Let the linear space $\mathcal{V}$ be given. For each $u \in \mathcal{V}$, the mapping $u \otimes : \mathbb{F} \to \mathcal{V}$, defined by the rule

$$u \otimes t := tu \quad \text{for all } t \in \mathbb{F},$$

is linear: it follows from (LS2) that this mapping is additive, and from (LS3) and the commutative law for multiplication in $\mathbb{F}$ that it is homogeneous. Note that $\text{Rng}(u\otimes) = \mathbb{F}u$, and that by Proposition 11C $u\otimes$ is injective if and only if $u \neq 0$. If $L : \mathcal{V} \to \mathcal{W}$ is a linear mapping, then $L(u\otimes) = (Lu)\otimes$ on account of the homogeneity of L, and we may omit the parentheses.

(h)*: We set $\text{Cont}(\mathbb{R}, \mathbb{R}) := \{f \in \text{Map}(\mathbb{R}, \mathbb{R}) \mid f \text{ is continuous}\}$ and $\text{Cont}^1(\mathbb{R}, \mathbb{R}) := \{f \in \text{Cont}(\mathbb{R}, \mathbb{R}) \mid f \text{ is differentiable, and its derivative } f^{\bullet} \text{ is continuous}\}$. These are linear spaces over $\mathbb{R}$, with the linear-space operations defined valuewise (i.e., they are *subspaces* of $\text{Map}(\mathbb{R}, \mathbb{R})$, in a sense to be made precise in Section 13). The mapping

$$f \mapsto f^{\bullet} : \text{Cont}^1(\mathbb{R}, \mathbb{R}) \to \text{Cont}(\mathbb{R}, \mathbb{R})$$

(*differentiation*) is linear. For given $a, b \in \mathbb{R}$, the mapping

$$f \mapsto \int_a^b f : \ \text{Cont}(\mathbb{R}, \mathbb{R}) \to \mathbb{R}$$

is linear. The mapping $J : \text{Cont}(\mathbb{R}, \mathbb{R}) \to \text{Map}(\mathbb{R}, \mathbb{R})$ defined by the rule

$$(Jf)(t) := \int_0^t f \quad \text{for all } t \in \mathbb{R} \text{ and } f \in \text{Cont}(\mathbb{R}, \mathbb{R})$$

is also linear. ■

To conclude this section, we note a few formulas.

12H. LEMMA. *Let the linear mapping* $L : \mathcal{V} \to \mathcal{W}$ *be given. Then*

$L_>(\mathcal{A} + \mathcal{B}) = L_>(\mathcal{A}) + L_>(\mathcal{B}) \quad \text{for all } \mathcal{A}, \mathcal{B} \in \mathfrak{P}(\mathcal{V})$

$L_>(K\mathcal{A}) = KL_>(\mathcal{A}) \quad \text{for all } K \in \mathfrak{P}(\mathbb{F}) \text{ and } \mathcal{A} \in \mathfrak{P}(\mathcal{V})$

$L^<(\mathcal{C} + \mathcal{D}) \supset L^<(\mathcal{C}) + L^<(\mathcal{D}) \quad \text{for all } \mathcal{C}, \mathcal{D} \in \mathfrak{P}(\mathcal{W})$

$L^<(K\mathcal{C}) = KL^<(\mathcal{C}) \quad \text{for all } K \in \mathfrak{P}(\mathbb{F}^\times) \text{ and } \mathcal{C} \in \mathfrak{P}(\mathcal{W}).$

Proof. Let $\mathcal{A}, \mathcal{B} \in \mathfrak{P}(\mathcal{V})$ and $K \in \mathfrak{P}(\mathbb{F})$ be given. Then the inclusions $L_>(\mathcal{A}+\mathcal{B}) \subset L_>(\mathcal{A}) + L_>(\mathcal{B})$ and $L_>(K\mathcal{A}) \subset KL_>(\mathcal{A})$ follow at once from (12.5) and (12.6). Let $w \in L_>(\mathcal{A}) + L_>(\mathcal{B})$ be given. We may choose $u \in L_>(\mathcal{A})$, $v \in L_>(\mathcal{B})$ such that $w = u + v$; we may further choose $a \in \mathcal{A}$, $b \in \mathcal{B}$ such that $u = La$, $v = Lb$. Then $w = u+v = La+Lb = L(a+b) \in L_>(\mathcal{A}+\mathcal{B})$. Since $w \in L_>(\mathcal{A})+L_>(\mathcal{B})$ was arbitrary, the reverse inclusion $L_>(\mathcal{A}) + L_>(\mathcal{B}) \subset L_>(\mathcal{A}+\mathcal{B})$ follows. A similar argument yields $KL_>(\mathcal{A}) \subset L_>(K\mathcal{A})$.

Let $\mathcal{C}, \mathcal{D} \in \mathfrak{P}(\mathcal{W})$ and $K \in \mathfrak{P}(\mathbb{F}^\times)$ be given. For all $u \in L^<(\mathcal{C})$ and $v \in L^<(\mathcal{D})$ we have $L(u + v) = Lu + Lv \in \mathcal{C} + \mathcal{D}$, and therefore $u + v \in L^<(\mathcal{C} + \mathcal{D})$. Thus $L^<(\mathcal{C}) + L^<(\mathcal{D}) \subset L^<(\mathcal{C} + \mathcal{D})$. For all $s \in K$ and $u \in L^<(\mathcal{C})$ we have $L(su) = sLu \in K\mathcal{C}$. Thus $KL^<(\mathcal{C}) \subset L^<(K\mathcal{C})$. Let $v \in L^<(K\mathcal{C})$ be given. We may choose $t \in K$ and $c \in \mathcal{C}$ such that $Lv = tc$. Then $L(\frac{1}{t}v) = \frac{1}{t}Lv = \frac{1}{t}tc = c \in \mathcal{C}$, so that $\frac{1}{t}v \in L^<(\mathcal{C})$ and $v = t\frac{1}{t}v \in KL^<(\mathcal{C})$. Thus $L^<(K\mathcal{C}) \subset KL^<(\mathcal{C})$. ■

13. Subspaces

Let the linear spaces $\mathcal{U}$ and $\mathcal{V}$ be given. Then $\mathcal{U}$ is called a (**linear**) **subspace of** $\mathcal{V}$ if $\mathcal{U} \subset \mathcal{V}$ and $1_{\mathcal{U}\subset\mathcal{V}}$ is linear.

Before we examine more closely the meaning of this definition, we establish some immediate consequences.

13A. PROPOSITION. *Let the linear spaces $\mathcal{V}$ and $\mathcal{W}$, the subspace $\mathcal{U}$ of $\mathcal{V}$, and the subspaces $\mathcal{X}$ and $\mathcal{Y}$ of $\mathcal{W}$ be given. For every linear mapping $L : \mathcal{V} \to \mathcal{X}$ such that $L_>(\mathcal{U}) \subset \mathcal{Y}$ the mapping $L|_{\mathcal{U}}^{\mathcal{Y}}$ obtained by adjustment is linear.*

Proof. Since $1_{\mathcal{U}\subset\mathcal{V}}$, L, and $1_{\mathcal{X}\subset\mathcal{W}}$ are linear, so is their composite $1_{\mathcal{X}\subset\mathcal{W}}\circ L\circ 1_{\mathcal{U}\subset\mathcal{V}} = L|_{\mathcal{U}}^{\mathcal{W}} = 1_{\mathcal{Y}\subset\mathcal{W}} \circ (L|_{\mathcal{U}}^{\mathcal{Y}})$, by Proposition 12C. Since $1_{\mathcal{Y}\subset\mathcal{W}}$ is linear and injective, $L|_{\mathcal{U}}^{\mathcal{Y}}$ is linear by Proposition 12D,(b). ∎

13B. COROLLARY. *Let the linear spaces $\mathcal{U}$, $\mathcal{V}$, $\mathcal{W}$ be given, and assume that $\mathcal{W}$ is a subspace of $\mathcal{V}$. Then $\mathcal{U}$ is a subspace of $\mathcal{W}$ if and only if $\mathcal{U} \subset \mathcal{W}$ and $\mathcal{U}$ is a subspace of $\mathcal{V}$.*

Proof. We may stipulate that $\mathcal{U} \subset \mathcal{W}$. We then have to prove that $1_{\mathcal{U}\subset\mathcal{W}} = 1_{\mathcal{U}\subset\mathcal{V}}|^{\mathcal{W}}$ is linear if and only if $1_{\mathcal{U}\subset\mathcal{V}} = 1_{\mathcal{U}\subset\mathcal{W}}|^{\mathcal{V}}$ is linear, and this follows from Proposition 13A. ∎

We return to the definition. It says that a subspace of the linear space $\mathcal{V}$ is a subset $\mathcal{U}$ of $\mathcal{V}$, endowed with the structure of a linear space in such a way that the operations of that structure "agree with" the corresponding operations of $\mathcal{V}$, in a sense to be made precise presently.

13C. LEMMA. *Let the linear space $\mathcal{V}$ and the subspace $\mathcal{U}$ of $\mathcal{V}$ be given. Then*

$$(13.1) \qquad 0^{\mathcal{V}} \in \mathcal{U}, \quad (\mathrm{add}^{\mathcal{V}})_>(\mathcal{U}\times\mathcal{U}) \subset \mathcal{U}, \quad (\mathrm{mult}^{\mathcal{V}}_s)_>(\mathcal{U}) \subset \mathcal{U} \quad \textit{for all } s \in \mathbb{F},$$

$$(13.2) \qquad 0^{\mathcal{U}} = 0^{\mathcal{V}}, \quad \mathrm{add}^{\mathcal{U}} = \mathrm{add}^{\mathcal{V}}|_{\mathcal{U}\times\mathcal{U}}^{\mathcal{U}}, \quad \mathrm{opp}^{\mathcal{U}} = \mathrm{opp}^{\mathcal{V}}|_{\mathcal{U}}^{\mathcal{U}},$$

$$\mathrm{mult}^{\mathcal{U}}_s = \mathrm{mult}^{\mathcal{V}}_s|_{\mathcal{U}}^{\mathcal{U}} \quad \textit{for all } s \in \mathbb{F}.$$

Proof. Since $1_{\mathcal{U}\subset\mathcal{V}}$ is linear, we have $0^{\mathcal{V}} = 1_{\mathcal{U}\subset\mathcal{V}}(0^{\mathcal{U}}) = 0^{\mathcal{U}}$; moreover,

$$\mathrm{add}^{\mathcal{V}}|_{\mathcal{U}\times\mathcal{U}} = \mathrm{add}^{\mathcal{V}} \circ 1_{\mathcal{U}\times\mathcal{U}\subset\mathcal{V}\times\mathcal{V}} = \mathrm{add}^{\mathcal{V}} \circ (1_{\mathcal{U}\subset\mathcal{V}} \times 1_{\mathcal{U}\subset\mathcal{V}}) = 1_{\mathcal{U}\subset\mathcal{V}} \circ \mathrm{add}^{\mathcal{U}}.$$

This shows that $(\mathrm{add}^{\mathcal{V}})_>(\mathcal{U}\times\mathcal{U}) = \mathrm{Rng}(\mathrm{add}^{\mathcal{V}}|_{\mathcal{U}\times\mathcal{U}}) \subset \mathcal{U}$ and that

$$\mathrm{add}^{\mathcal{V}}|_{\mathcal{U}\times\mathcal{U}}^{\mathcal{U}} = (1_{\mathcal{U}\subset\mathcal{V}}|^{\mathcal{U}}) \circ \mathrm{add}^{\mathcal{U}} = 1_{\mathcal{U}} \circ \mathrm{add}^{\mathcal{U}} = \mathrm{add}^{\mathcal{U}}.$$

The proof for the scalar multiplication is similar, but simpler. The proof for the opposition follows by (11.7). ∎

Lemma 13C shows that a subset $\mathcal{U}$ of a linear space $\mathcal{V}$ can be endowed with at most one linear-space structure so as to be a subspace of $\mathcal{V}$, and the conditions in (13.1) are necessary for such a structure to exist. We shall now show that they are also sufficient.

13D. Theorem. *Let the linear space* $\mathcal{V}$ *and the subset* $\mathcal{U}$ *of* $\mathcal{V}$ *be given. Then* $\mathcal{U}$ *can be endowed with the structure of a linear subspace of* $\mathcal{V}$ *if and only if*

$$(13.3) \qquad\qquad 0 \in \mathcal{U}, \qquad \mathcal{U} + \mathcal{U} \subset \mathcal{U}, \qquad \mathbb{F}\mathcal{U} \subset \mathcal{U}.$$

When these conditions (restatements of (13.1)) are satisfied, there is exactly one such structure, and it is defined by requiring (13.2).

Proof. The "only if" part follows from Lemma 13C, as does the fact that any subspace structure of $\mathcal{U}$ must be given by (13.2).

To prove the "if" part, we assume that (13.3) holds. Then (13.1) holds, and we may define $0^{\mathcal{U}}$, $\mathrm{add}^{\mathcal{U}}$, $\mathrm{opp}^{\mathcal{U}}$, $\mathrm{mult}^{\mathcal{U}}$ by requiring that (13.2) hold. It remains to verify that this prescription satisfies the defining laws for a linear space, and that $1_{\mathcal{U} \subset \mathcal{V}}$ is linear. The validity of each of these laws (the associative, commutative, and neutrality laws for addition, the law of opposites, and (LS1)-(LS4)) follows at once from the validity of the corresponding law for $\mathcal{V}$, since it expresses the holding of *the same equality* when the objects to be tested are members of the *subset* $\mathcal{U}$ of $\mathcal{V}$. The fact that $1_{\mathcal{U} \subset \mathcal{V}}$ is linear is verified as follows:

$$\begin{aligned}
1_{\mathcal{U} \subset \mathcal{V}} \circ \mathrm{add}^{\mathcal{U}} &= 1_{\mathcal{U} \subset \mathcal{V}} \circ (\mathrm{add}^{\mathcal{V}}|_{\mathcal{U} \times \mathcal{U}}^{\mathcal{U}}) = \mathrm{add}^{\mathcal{V}}|_{\mathcal{U} \times \mathcal{U}} = \mathrm{add}^{\mathcal{V}} \circ 1_{\mathcal{U} \times \mathcal{U} \subset \mathcal{V} \times \mathcal{V}} = \\
&= \mathrm{add}^{\mathcal{V}} \circ (1_{\mathcal{U} \subset \mathcal{V}} \times 1_{\mathcal{U} \subset \mathcal{V}}) \\
1_{\mathcal{U} \subset \mathcal{V}} \circ \mathrm{mult}^{\mathcal{U}}{}_s &= 1_{\mathcal{U} \subset \mathcal{V}} \circ (\mathrm{mult}^{\mathcal{V}}{}_s|_{\mathcal{U}}^{\mathcal{U}}) = \\
&= \mathrm{mult}^{\mathcal{V}}{}_s|_{\mathcal{U}} = \mathrm{mult}^{\mathcal{V}}{}_s \circ 1_{\mathcal{U} \subset \mathcal{V}} \quad \text{for all } s \in \mathbb{F}. \ \blacksquare
\end{aligned}$$

13E. Remarks. (a): In view of the uniqueness of the subspace structure on $\mathcal{U}$, we claim the license to call the *subset* $\mathcal{U}$ of $\mathcal{V}$ a **subspace of** $\mathcal{V}$ when the conditions (13.3) are satisfied, and to then regard $\mathcal{U}$, without explicit mention, as endowed with the structure of a linear space by prescribing the operations according to (13.2).

(b): The empty subset $\varnothing$ of $\mathcal{V}$ satisfies $\varnothing + \varnothing \subset \varnothing$ and $\mathbb{F}\varnothing \subset \varnothing$, but it is not a subspace. The condition $0 \in \mathcal{U}$ in (13.1) is therefore not redundant. It may, however, be replaced by the apparently weaker condition $\mathcal{U} \neq \varnothing$; indeed, after this replacement, we may choose $a \in \mathcal{U}$ and find $0 = 0a \in \mathbb{F}\mathcal{U} \subset \mathcal{U}$.

(c): If $\mathcal{U}$ is a subspace of $\mathcal{V}$ we have $\mathcal{U} = 0 + \mathcal{U} \subset \mathcal{U} + \mathcal{U}$ and $\mathcal{U} = s(\frac{1}{s}\mathcal{U}) \subset s\mathcal{U}$ for every $s \in \mathbb{F}^{\times}$, so that (13.3) may be sharpened to

$$0 \in \mathcal{U}, \qquad \mathcal{U} + \mathcal{U} = \mathcal{U}, \qquad s\mathcal{U} = \mathcal{U} \text{ for every } s \in \mathbb{F}^{\times}.$$

(d): Let the linear space $\mathcal{V}$ and the subspace $\mathcal{U}$ of $\mathcal{V}$ be given. We have used the same symbol $+$ for addition in $\mathcal{V}$ and in $\mathcal{U}$, and there is no clash between these uses because of Lemma 13C. We generalize this remark. Let a family $z \in \mathcal{U}^I$ be given. For every finite subset J of I, and for every subset J of I if $\mathrm{Supp}z$ is finite (this is unambiguous, since $0^{\mathcal{U}} = 0^{\mathcal{V}}$), there are two possible interpretations of $\sum_J z$, one in the commutative monoid $\mathcal{U}$ with addition $\mathrm{add}^{\mathcal{U}}$, the other in the commutative monoid $\mathcal{V}$ with addition $\mathrm{add}^{\mathcal{V}}$. It follows by special induction from (13.2) that these interpretations of $\sum_J z$ agree if J is finite; and therefore, if $\mathrm{Supp} z$ is finite, both interpretations of $\sum_J z = \sum_{J \cap \mathrm{Supp} z} z$ agree for *every* subset J of I. We may therefore write $\sum_J z$ without ambiguity, and moreover obtain the following generalization of the formula $\mathcal{U} + \mathcal{U} \subset \mathcal{U}$ in (13.3). ∎

13F. COROLLARY. *Let the linear space* $\mathcal{V}$, *the subspace* $\mathcal{U}$ *of* $\mathcal{V}$, *and the index set* I *be given. Then* $\sum_{i \in I} \mathcal{U} \subset \mathcal{U}$.

13G. EXAMPLES. (a): Let the linear space $\mathcal{V}$ be given. Then $\{0\}$ and $\mathcal{V}$ are subspaces of $\mathcal{V}$. They are called the **zero-subspace**, or **trivial subspace, of** $\mathcal{V}$, and the **improper subspace of** $\mathcal{V}$, respectively. All subspaces of $\mathcal{V}$ other than $\mathcal{V}$ itself are proper subsets of $\mathcal{V}$, and are therefore called **proper subspaces** of $\mathcal{V}$.

(b): Let the linear space $\mathcal{V}$ be given. If $\mathcal{U}_1$, $\mathcal{U}_2$ are subspaces of $\mathcal{V}$, then $\mathcal{U}_1 \cap \mathcal{U}_2$ and $\mathcal{U}_1 + \mathcal{U}_2$ are also subspaces of $\mathcal{V}$, as is readily verified by using Theorem 13D. These are special cases of results involving an arbitrary family of subspaces (Propositions 13J and 13R).

(c): Let the linear spaces $\mathcal{V}$ and $\mathcal{W}$ be given. A mapping $L : \mathcal{V} \to \mathcal{W}$ is linear if and only if its graph $\mathrm{Gr}(L)$ is a subspace of the linear space $\mathcal{V} \times \mathcal{W}$ (Example 11D,(c)).

(d): Let the families of linear spaces $(\mathcal{U}_i \mid i \in I)$ and $(\mathcal{V}_i \mid i \in I)$ be given, and assume that $\mathcal{U}_i$ is a subspace of $\mathcal{V}_i$ for every $i \in I$. Then $\bigtimes_{i \in I} \mathcal{U}_i$ is a subspace of $\bigtimes_{i \in I} \mathcal{V}_i$. In particular, if $\mathcal{V}$ is a linear space and $\mathcal{U}$ is a subspace of $\mathcal{V}$, then $\mathcal{U}^I$ is a subspace of $\mathcal{V}^I$ for every set I.

(e): Let the family of linear spaces $(\mathcal{V}_i \mid i \in I)$ be given. For all $a, b \in \bigtimes_{i \in I} \mathcal{V}_i$ and $s \in \mathbb{F}$ we have $\mathrm{Supp}(a+b) \subset \mathrm{Supp} a \cup \mathrm{Supp} b$ and $\mathrm{Supp}(sa) \subset \mathrm{Supp} a$. Therefore the subset

$$\bigoplus_{i \in I} \mathcal{V}_i := \{a \in \bigtimes_{i \in I} \mathcal{V}_i \mid \mathrm{Supp} a \text{ is finite}\}$$

of $\bigtimes_{i \in I} \mathcal{V}_i$ is a subspace of $\bigtimes_{i \in I} \mathcal{V}_i$. It is called the **direct sum of** the family $(\mathcal{V}_i \mid, i \in I)$. We observe that $\bigoplus_{i \in I} \mathcal{V}_i = \bigtimes_{i \in I} \mathcal{V}_i$ if and •only if $\mathrm{Supp}(\mathcal{V}_i \mid i \in I)$ is finite; this equality therefore holds if I is finite.

In particular, let the set I and the linear space $\mathcal{V}$ be given. Then (cf. *Basic Language*, Section 117) $\mathcal{V}^{(I)} := \{z \in \mathcal{V}^I \mid \mathrm{Supp} z \text{ is finite}\} = \bigoplus_{i \in I} \mathcal{V}$ is a subspace of $\mathcal{V}^I$. For *every* subset J of I the mapping

$$a \mapsto \sum_J a : \mathcal{V}^{(I)} \to \mathcal{V}$$

is linear (the proof is as in Example 12G,(f)). We note that $\mathcal{V}^{(I)} = \mathcal{V}^I$ if and only if I is finite or $\mathcal{V}$ is a zero-space.

More in particular, $\mathbb{F}^{(I)}$ contains the Kronecker family δ_i^I for each $i \in I$ (here we take the terms 0 and 1 of the Kronecker families to be the zero and unity of the field $\mathbb{F}$). Let $a \in \mathbb{F}^{(I)}$ be given. Then $\mathrm{Supp}(a_i \delta_i^I \mid i \in I) = \mathrm{Supp}a$ is finite, and hence, by Proposition 12B and Example 12G,(a),

$$(\sum_{i \in I} a_i \delta_i^I)_j = \pi_j \sum_{i \in I} a_i \delta_i^I = \sum_{i \in I} \pi_j(a_i \delta_i^I) = \sum_{i \in I} a_i \delta_{i,j}^I = a_j \quad \text{for all } j \in I.$$

We thus see that

(13.4) $$a = \sum_{i \in I} a_i \delta_i^I \quad \text{for all } a \in \mathbb{F}^{(I)}$$

(f)*: Let the subset S of $\mathbb{R}$ be given. Then $\mathrm{Cont}(S, \mathbb{R}) := \{f \in \mathrm{Map}(S, \mathbb{R}) \mid f$ is continuous$\}$ is a subspace of $\mathrm{Map}(S, \mathbb{R})$. Moreover, $\mathrm{Cont}^1(\mathbb{R}, \mathbb{R}) := \{f \in \mathrm{Cont}(\mathbb{R}, \mathbb{R}) \mid f$ is differentiable, and its derivative $f^\cdot$ is continuous$\}$ is a subspace of $\mathrm{Cont}(\mathbb{R}, \mathbb{R})$. (These are linear spaces over $\mathbb{R}$.) ■

Particularly important instances of subspaces are the sets $\mathrm{Lin}(\mathcal{V}, \mathcal{W})$ of linear mappings.

13H. PROPOSITION. *Let the linear spaces $\mathcal{V}$ and $\mathcal{W}$ be given. Then $\mathrm{Lin}(\mathcal{V}, \mathcal{W})$ is a subspace of $\mathrm{Map}(\mathcal{V}, \mathcal{W})$.*

Proof. $0_{\mathcal{V} \to \mathcal{W}} \in \mathrm{Lin}(\mathcal{V}, \mathcal{W})$ (Remark 12A,(c)). Let $L, M \in \mathrm{Lin}(\mathcal{V}, \mathcal{W})$ and $t \in \mathbb{F}$ be given. Then

$$\forall u, v \in \mathcal{V}, \quad (L + M)(u + v) = L(u + v) + M(u + v) =$$
$$= (Lu + Lv) + (Mu + Mv) =$$
$$= (Lu + Mu) + (Lv + Mv) =$$
$$= (L + M)(u) + (L + M)(v)$$

$$\forall s \in \mathbb{F}, \forall u \in \mathcal{V}, \quad (L + M)(su) = L(su) + M(su) = sLu + sMu =$$
$$= s(Lu + Mu) = s(L + M)(u);$$

hence $L + M \in \mathrm{Lin}(\mathcal{V}, \mathcal{W})$; and

$$\forall u, v \in \mathcal{V}, \quad (tL)(u + v) = tL(u + v) = t(Lu + Lv) = tLu + tLv =$$
$$= (tL)(u) + (tL)(v)$$

$$\forall s \in \mathbb{F}, \forall u \in \mathcal{V}, \quad (tL)(su) = tL(su) = t(sLu) = (ts)Lu =$$
$$= (st)Lu = s(tLu) = s(tL)(u);$$

hence $tL \in \mathrm{Lin}(\mathcal{V}, \mathcal{W})$. Since $L, M \in \mathrm{Lin}(\mathcal{V}, \mathcal{W})$ and $t \in \mathbb{F}$ were arbitrary, it follows from Theorem 13D that $\mathrm{Lin}(\mathcal{V}, \mathcal{W})$ is a subspace of $\mathrm{Map}(\mathcal{V}, \mathcal{W})$. ■

13I. REMARK. Let the linear spaces $\mathcal{V}$, $\mathcal{W}$, $\mathcal{X}$ be given. Then Proposition 13H and Examples 12G,(d),(e) show that for all $K, L \in \mathrm{Lin}(\mathcal{V}, \mathcal{W})$, $M, N \in \mathrm{Lin}(\mathcal{W}, \mathcal{X})$, and $s, t \in \mathbb{F}$ we have

$$M(K+L) = MK+ML, \qquad (M+N)K = MK+NK, \qquad (sM)(tK) = (st)(MK).$$

Therefore the mappings $(L \mapsto ML) : \mathrm{Lin}(\mathcal{V}, \mathcal{W}) \to \mathrm{Lin}(\mathcal{V}, \mathcal{X})$ and $(N \mapsto NK) : \mathrm{Lin}(\mathcal{W}, \mathcal{X}) \to \mathrm{Lin}(\mathcal{V}, \mathcal{X})$ are linear. ∎

For each linear space $\mathcal{V}$ we define

$$\mathrm{Subsp}(\mathcal{V}) := \{\mathcal{U} \in \mathfrak{P}(\mathcal{V}) \mid \mathcal{U} \text{ is a subspace of } \mathcal{V}\}.$$

13J. PROPOSITION. *Let the linear space $\mathcal{V}$ be given. Then* $\mathrm{Subsp}(\mathcal{V})$ *is an intersection-stable subcollection of* $\mathfrak{P}(\mathcal{V})$.

Proof. By Theorem 13D we have

$$\mathrm{Subsp}(\mathcal{V}) = \{\mathcal{U} \in \mathfrak{P}(\mathcal{V}) \mid 0 \in \mathcal{U},\ \mathcal{U} + \mathcal{U} \subset \mathcal{U},\ \mathbb{F}\mathcal{U} \subset \mathcal{U}\}.$$

Let Γ be a subcollection of $\mathrm{Subsp}(\mathcal{V})$. Then $0 \in \mathcal{U}$ for every $\mathcal{U} \in \Gamma$, and therefore $0 \in \bigcap^{\vee}\Gamma$. We also have $\bigcap^{\vee}\Gamma + \bigcap^{\vee}\Gamma \subset \mathcal{U} + \mathcal{U} \subset \mathcal{U}$ and $\mathbb{F}(\bigcap^{\vee}\Gamma) \subset \mathbb{F}\mathcal{U} \subset \mathcal{U}$ for all $\mathcal{U} \in \Gamma$, and therefore $\bigcap^{\vee}\Gamma + \bigcap^{\vee}\Gamma \subset \bigcap^{\vee}\Gamma$ and $\mathbb{F}(\bigcap^{\vee}\Gamma) \subset \bigcap^{\vee}\Gamma$. It follows that $\bigcap^{\vee}\Gamma \in \mathrm{Subsp}(\mathcal{V})$. ∎

13K. COROLLARY. *Let the linear space $\mathcal{V}$ and the subspace $\mathcal{W}$ of $\mathcal{V}$ be given. Then*

$$\mathrm{Subsp}(\mathcal{W}) = \mathrm{Subsp}(\mathcal{V}) \cap \mathfrak{P}(\mathcal{W}) = \{\mathcal{W} \cap \mathcal{U} \mid \mathcal{U} \in \mathrm{Subsp}(\mathcal{V})\}.$$

Proof. The first equality is a restatement of Corollary 13B. For every $\mathcal{U} \in \mathrm{Subsp}(\mathcal{W})$ we have $\mathcal{W} \cap \mathcal{U} = \mathcal{U}$, and $\mathcal{U} \in \mathrm{Subsp}(\mathcal{V})$ by Corollary 13B. Conversely, if $\mathcal{U} \in \mathrm{Subsp}(\mathcal{V})$, then $\mathcal{W} \cap \mathcal{U} \in \mathrm{Subsp}(\mathcal{V}) \cap \mathfrak{P}(\mathcal{W})$ by Proposition 13J (or Example 13G,(b)). ∎

According to *Basic Language*, Theorem 73D,(b) and Remark 73E, the intersection-stable subcollection $\mathrm{Subsp}(\mathcal{V})$ of $\mathfrak{P}(\mathcal{V})$ is the range (and hence also the set of fixed points) of exactly one closure mapping in $\mathfrak{P}(\mathcal{V})$ ordered by inclusion. We denote this closure mapping by $\mathrm{Lsp}^{\mathcal{V}}$; it is given by the rule

$$(13.5) \qquad \mathrm{Lsp}^{\mathcal{V}}\mathcal{A} := \bigcap\{\mathcal{U} \in \mathrm{Subsp}(\mathcal{V}) \mid \mathcal{A} \subset \mathcal{U}\} \quad \text{for all } \mathcal{A} \in \mathfrak{P}(\mathcal{V})$$

(the collection of which the intersection is taken contains $\mathcal{V}$, and is therefore not empty). We call $\mathrm{Lsp}^{\mathcal{V}}\mathcal{A}$ the **linear span (in $\mathcal{V}$) of** $\mathcal{A}$; it is the smallest among the subspaces of $\mathcal{V}$ that include $\mathcal{A}$. The next proposition shows that in most contexts the index $\mathcal{V}$ and the phrase "in $\mathcal{V}$" may be omitted, so that we shall usually write $\mathrm{Lsp}\mathcal{A}$.

13L. Proposition. *Let the linear space $\mathcal{V}$ and the subspace $\mathcal{W}$ of $\mathcal{V}$ be given. Then $(\mathrm{Lsp}^{\mathcal{V}})_{>}(\mathfrak{P}(\mathcal{W})) \subset \mathfrak{P}(\mathcal{W})$, and*

$$\mathrm{Lsp}^{\mathcal{W}} = \mathrm{Lsp}^{\mathcal{V}}\big|_{\mathfrak{P}(\mathcal{W})}^{\mathfrak{P}(\mathcal{W})},$$

and $\mathrm{Lsp}^{\mathcal{W}}\mathcal{A} = \mathrm{Lsp}^{\mathcal{V}}\mathcal{A}$ for all $\mathcal{A} \in \mathfrak{P}(\mathcal{W})$.

Proof. Let $\mathcal{A} \in \mathfrak{P}(\mathcal{W})$ be given. Then $\mathcal{A} \subset \mathcal{W} \in \mathrm{Subsp}(\mathcal{V})$, and hence $\mathrm{Lsp}^{\mathcal{V}}\mathcal{A} \subset \mathcal{W}$. By Corollary 13B, $\mathcal{A} \subset \mathrm{Lsp}^{\mathcal{V}}\mathcal{A} \in \mathrm{Subsp}(\mathcal{W})$, and hence $\mathrm{Lsp}^{\mathcal{W}}\mathcal{A} \subset \mathrm{Lsp}^{\mathcal{V}}\mathcal{A}$. On the other hand, $\mathcal{A} \subset \mathrm{Lsp}^{\mathcal{W}}\mathcal{A} \in \mathrm{Subsp}(\mathcal{W}) \subset \mathrm{Subsp}(\mathcal{V})$; therefore $\mathrm{Lsp}^{\mathcal{V}}\mathcal{A} \subset \mathrm{Lsp}^{\mathcal{W}}\mathcal{A}$. ∎

13M. Proposition. *Let the linear space $\mathcal{V}$ be given. The collection $\mathrm{Subsp}(\mathcal{V})$ is completely ordered by inclusion. For every subcollection Γ of $\mathrm{Subsp}(\mathcal{V})$ we have $\inf_{\mathrm{Subsp}(\mathcal{V})}\Gamma = \bigcap^{\mathcal{V}}\Gamma$ and $\sup_{\mathrm{Subsp}(\mathcal{V})}\Gamma = \mathrm{Lsp}\bigcup\Gamma$.*

Proof. This follows immediately from Proposition 13J and *Basic Language*, Propositions 71F and 73I,(b). ∎

Later, in Corollary 13R, we shall give another formula for the supremum.

We observe that the union of a collection of subspaces is in general not a subspace: indeed, if $\mathcal{U}_1$, $\mathcal{U}_2$ are subspaces of the linear space $\mathcal{V}$, $\mathcal{U}_1 \cup \mathcal{U}_2$ is a subspace if and only if either $\mathcal{U}_1 \subset \mathcal{U}_2$ or $\mathcal{U}_2 \subset \mathcal{U}_1$, as follows easily from Theorem 13D. There is, however, a special kind of collection of subspaces for which the union actually is a subspace.

13N. Proposition. *Let the linear space $\mathcal{V}$ and the non-empty subcollection Γ of $\mathrm{Subsp}(\mathcal{V})$ be given. If Γ is directed by inclusion, and in particular if Γ is a nest, then $\bigcup\Gamma \in \mathrm{Subsp}(\mathcal{V})$.*

Proof. It is obvious from Theorem 13D, without any assumption on the non-empty subcollection Γ of $\mathrm{Subsp}(\mathcal{V})$, that $0 \in \bigcup\Gamma$ and $\mathbb{F}(\bigcup\Gamma) \subset \bigcup\Gamma$. Now suppose that Γ is directed by inclusion. Let $u, w \in \bigcup\Gamma$ be given. We may choose $\mathcal{U}, \mathcal{W} \in \Gamma$ such that $u \in \mathcal{U}$, $w \in \mathcal{W}$. We may further choose $\mathcal{X} \in \Gamma$ such that $\mathcal{U} \subset \mathcal{X}$, $\mathcal{W} \subset \mathcal{X}$. Then $u + w \in \mathcal{U} + \mathcal{W} \subset \mathcal{X} + \mathcal{X} \subset \mathcal{X} \subset \bigcup\Gamma$. Since $u, w \in \bigcup\Gamma$ were arbitrary, it follows that $\bigcup\Gamma + \bigcup\Gamma \subset \bigcup\Gamma$. By Theorem 13D, $\bigcup\Gamma \in \mathrm{Subsp}(\mathcal{V})$. ∎

13O. Corollary. *Let the linear space $\mathcal{V}$ be given. Then*

$$\mathrm{Lsp}\mathcal{A} = \bigcup_{\mathcal{B} \in \mathfrak{F}(\mathcal{A})} \mathrm{Lsp}\mathcal{B} \quad for\ all\ \mathcal{A} \in \mathfrak{P}(\mathcal{V}).$$

Proof. Since $\mathcal{B} \subset \mathcal{A}$, and hence $\mathrm{Lsp}\mathcal{B} \subset \mathrm{Lsp}\mathcal{A}$, for all $\mathcal{B} \in \mathfrak{F}(\mathcal{A})$, we have $\mathrm{Lsp}\mathcal{A} \supset \bigcup_{\mathcal{B} \in \mathfrak{F}(\mathcal{A})} \mathrm{Lsp}\mathcal{B}$.

If $\mathcal{B}, \mathcal{B}' \in \mathfrak{F}(\mathcal{A})$, then $\mathcal{B} \cup \mathcal{B}' \in \mathfrak{F}(\mathcal{A})$ and $\mathrm{Lsp}\mathcal{B} \cup \mathrm{Lsp}\mathcal{B}' \subset \mathrm{Lsp}(\mathcal{B} \cup \mathcal{B}')$. This shows that the subcollection $\{\mathrm{Lsp}\mathcal{B} \mid \mathcal{B} \in \mathfrak{F}(\mathcal{A})\}$ of $\mathrm{Subsp}(\mathcal{V})$ is directed by inclusion. by Proposition 13N, $\bigcup_{\mathcal{B} \in \mathfrak{F}(\mathcal{A})} \mathrm{Lsp}\mathcal{B}$ is a subspace of $\mathcal{V}$. Since $\mathcal{A} = \bigcup_{\mathcal{B} \in \mathfrak{F}(\mathcal{A})} \mathcal{B} \subset \bigcup_{\mathcal{B} \in \mathfrak{F}(\mathcal{A})} \mathrm{Lsp}\mathcal{B}$, we conclude that $\mathrm{Lsp}\mathcal{A} \subset \bigcup_{\mathcal{B} \in \mathfrak{F}(\mathcal{A})} \mathrm{Lsp}\mathcal{B}$. ∎

We now turn to the interaction between subspaces and linear mappings.

13P. Theorem. *Let the linear spaces V and W and the linear mapping $L : V \to W$ be given. Then*

$$(13.6) \qquad\qquad (L_>)_> (\mathrm{Subsp}(V)) \subset \mathrm{Subsp}(W)$$

$$(13.7) \qquad\qquad (L^<)_> (\mathrm{Subsp}(W)) \subset \mathrm{Subsp}(V).$$

In particular,

$$\mathrm{Rng}\, L = L_>(V) \in \mathrm{Subsp}(W)$$
$$L^<(\{0\}) \in \mathrm{Subsp}(V).$$

Proof. Let $\mathcal{U} \in \mathrm{Subsp}(V)$ be given. Then $0^W = L(0^V) \in L_>(\mathcal{U})$. Moreover, Lemma 12H and Theorem 13D yield $L_>(\mathcal{U}) + L_>(\mathcal{U}) = L_>(\mathcal{U} + \mathcal{U}) \subset L_>(\mathcal{U})$ and $\mathbb{F} L_>(\mathcal{U}) = L_>(\mathbb{F}\mathcal{U}) \subset L_>(\mathcal{U})$. By Theorem 13D again, it follows that $L_>(\mathcal{U}) \in \mathrm{Subsp}(W)$. This proves (13.6).

Let $\mathcal{X} \in \mathrm{Subsp}(W)$ be given. By Lemma 12H and Theorem 13D we have $L^<(\mathcal{X}) + L^<(\mathcal{X}) \subset L^<(\mathcal{X} + \mathcal{X}) \subset L^<(\mathcal{X})$, $\mathbb{F}^\times L^<(\mathcal{X}) = L^<(\mathbb{F}^\times \mathcal{X}) \subset L^<(\mathcal{X})$, and $0^V \in 0L^<(\mathcal{X}) = \{0^V\} \subset L^<(\mathcal{X})$. By Theorem 13D again, it follows that $L^<(\mathcal{X}) \in \mathrm{Subsp}(V)$. This proves (13.7). ∎

If L is a linear mapping, Theorem 13P shows that $L^<(\{0\})$ is a subspace of $\mathrm{Dom}\, L$. This subspace is called the **null-space of** L and is denoted by $\mathrm{Null}\, L$. (Sometimes the term *kernel of* L and the notation $\mathrm{Ker}\, L$ are used instead.) We observe that, if L and M are linear mappings with $\mathrm{Dom}\, M = \mathrm{Cod}\, L$, we have

$$(13.8) \qquad\qquad \mathrm{Null}(ML) = L^<(\mathrm{Null}\, M) \supset \mathrm{Null}\, L.$$

13Q. Corollary. *Let the linear spaces V and W and the linear mapping $L : V \to W$ be given. Then $L_> \circ \mathrm{Lsp}^V = \mathrm{Lsp}^W \circ L_>$.*

Proof. Let $\mathcal{A} \in \mathfrak{P}(V)$ be given. By Theorem 13P, $L_>(\mathrm{Lsp}^V \mathcal{A}) \in \mathrm{Subsp}(W)$; but since $\mathcal{A} \subset \mathrm{Lsp}^V \mathcal{A}$, we have $L_>(\mathcal{A}) \subset L_>(\mathrm{Lsp}^V \mathcal{A})$. It follows that

$$(13.9) \qquad\qquad \mathrm{Lsp}^W L_>(\mathcal{A}) \subset L_>(\mathrm{Lsp}^V \mathcal{A}).$$

On the other hand, $L^<(\mathrm{Lsp}^W L_>(\mathcal{A})) \in \mathrm{Subsp}(V)$, by Theorem 13P; but $L_>(\mathcal{A}) \subset \mathrm{Lsp}^W L_>(\mathcal{A})$, and therefore $\mathcal{A} \subset L^<(L_>(\mathcal{A})) \subset L^<(\mathrm{Lsp}^W L_>(\mathcal{A}))$. It follows that $\mathrm{Lsp}^V \mathcal{A} \subset L^<(\mathrm{Lsp}^W L_>(\mathcal{A}))$, and therefore

$$(13.10) \qquad L_>(\mathrm{Lsp}^V \mathcal{A}) \subset L_>(L^<(\mathrm{Lsp}^W L_>(\mathcal{A}))) \subset \mathrm{Lsp}^W L_>(\mathcal{A}).$$

Combination (13.9) and (13.10) yields $L_>(\mathrm{Lsp}^V \mathcal{A}) = \mathrm{Lsp}^W(\mathcal{A})$. But $\mathcal{A} \in \mathfrak{P}(V)$ was arbitrary; the assertion follows. ∎

13R. Proposition. *Let the linear space $\mathcal{V}$ be given. For every family $(\mathcal{U}_i \mid i \in I)$* in $\mathrm{Subsp}(\mathcal{V})$ *we have*

$$\sum_{i\in I}\mathcal{U}_i = \sup{}_{\mathrm{Subsp}(\mathcal{V})}\{\mathcal{U}_i \mid i \in I\} = \mathrm{Lsp}\bigcup_{i\in I}\mathcal{U}_i \in \mathrm{Subsp}(\mathcal{V}).$$

Proof. We recall that $\mathcal{V}^{(I)}$ is a subspace of $\mathcal{V}^I$ and that the mapping

$$(13.11) \qquad\qquad a \mapsto \sum_J a : \mathcal{V}^{(I)} \to \mathcal{V}$$

is linear (Example 13G,(e)). Moreover, $\underset{i\in I}{\bigtimes}\mathcal{U}_i$ is a subspace of $\mathcal{V}^I$ (Example 13G,(d)), and therefore $\mathcal{V}^{(I)} \cap \underset{i\in I}{\bigtimes}\mathcal{U}_i$ is a subspace of $\mathcal{V}^{(I)}$ (Corollary 13K). Now $\sum_{i\in I}\mathcal{U}_i$ (see definition in *Basic Language*, Section 131) is precisely the image of this subspace under the linear mapping given by (13.11). By Theorem 13P, $\sum_{i\in I}\mathcal{U}_i$ is a subspace of $\mathcal{V}$.

Set $\mathcal{W} := \sup_{\mathrm{Subsp}(\mathcal{V})}\{\mathcal{U}_i \mid i \in I\} = \mathrm{Lsp}\bigcup_{i\in I}\mathcal{U}_i$ (cf. Proposition 13M). We have $\underset{i\in I}{\bigtimes}\mathcal{U}_i \subset \underset{i\in I}{\bigtimes}\mathcal{W}$ and therefore, using Corollary 13F, $\sum_{i\in I}\mathcal{U}_i \subset \sum_{i\in I}\mathcal{W} \subset \mathcal{W}$.

On the other hand, let $j \in I$ be given, and define the family $(\mathcal{T}_i \mid i \in I)$ in $\mathrm{Subsp}(\mathcal{V})$ by the rule

$$\mathcal{T}_i := \begin{cases} \mathcal{U}_j & \text{if } i = j \\ \{0\} & \text{if } i \in I\backslash\{j\}. \end{cases}$$

Then $\underset{i\in I}{\bigtimes}\mathcal{T}_i \subset \underset{i\in I}{\bigtimes}\mathcal{U}_i$, and therefore $\mathcal{U}_j = \sum_{i\in I}\mathcal{T}_i \subset \sum_{i\in I}\mathcal{U}_i$. Since $j \in I$ was arbitrary, we have $\bigcup_{i\in I}\mathcal{U}_i \subset \sum_{i\in I}\mathcal{U}_i$. Since $\sum_{i\in I}\mathcal{U}_i \in \mathrm{Subsp}(\mathcal{V})$, we have $\mathcal{W} = \mathrm{Lsp}\bigcup_{i\in I}\mathcal{U}_i \subset \sum_{i\in I}\mathcal{U}_i$. We conclude that $\mathcal{W} = \sum_{i\in I}\mathcal{U}_i$. ∎

13S. Corollary. *Let the linear space $\mathcal{V}$ be given. For every family $(\mathcal{A}_i \mid i \in I)$* in $\mathfrak{P}(\mathcal{V})$ *we have*

$$\mathrm{Lsp}\bigcup_{i\in I}\mathcal{A}_i = \sum_{i\in I}\mathrm{Lsp}\mathcal{A}_i.$$

Proof. We apply *Basic Language*, Proposition 73I,(a) to the closure mapping $\mathrm{Lsp}^\mathcal{V}$ and find

$$\mathrm{Lsp}\bigcup_{i\in I}\mathcal{A}_i = \sup{}_{\mathrm{Subsp}(\mathcal{V})}\{\mathrm{Lsp}\mathcal{A}_i \mid i \in I\} = \sum_{i\in I}\mathrm{Lsp}\mathcal{A}_i,$$

where the second equality follows from Proposition 13R. ∎

13T. Proposition. *Let the linear space $\mathcal{V}$ be given. We have* $\mathrm{Lsp}\emptyset = \{0\}$,

(13.12) $$\mathrm{Lsp}\{u\} = \mathbb{F}u \quad for\ all\ u \in \mathcal{V}$$

(13.13) $$\mathrm{Lsp}\mathcal{A} = \sum_{u \in \mathcal{A}} \mathbb{F}u \quad for\ all\ \mathcal{A} \in \mathfrak{P}(\mathcal{V}).$$

Proof. The equality $\mathrm{Lsp}\emptyset = \{0\}$ is trivial. Let $u \in \mathcal{V}$ be given. By Theorem 13D, $\mathbb{F}u \subset \mathbb{F}\mathrm{Lsp}\{u\} \subset \mathrm{Lsp}\{u\}$. But $0 = 0u \in \mathbb{F}u, \mathbb{F}u + \mathbb{F}u = (\mathbb{F} + \mathbb{F})u = \mathbb{F}u, \mathbb{F}(\mathbb{F}u) = (\mathbb{F}\mathbb{F})u = \mathbb{F}u$, so that $\mathbb{F}u$ is a subspace of $\mathcal{V}$, and contains $u = 1u$; and hence $\mathrm{Lsp}\{u\} = \mathbb{F}u$. This establishes (13.12). (13.13) then follows from (13.12) and Corollary 13S. ∎

14. Linear partitions

14A. PROPOSITION. *Let the linear space V and the subspace U of V be given. Then*

$$V/U := \{v + U \mid v \in V\}$$

is a partition of V, and its partition mapping $\Omega_{V/U} : V \to V/U$ satisfies

$$\Omega_{V/U}(v) = v + U \quad \text{for all } v \in V.$$

Proof. We define the mapping $\Omega \colon V \to V/U$ by the rule

$$\Omega(v) := v + U \quad \text{for all } v \in V,$$

and claim that

(14.1) $\Omega^<(\{\Omega(v)\}) = v + U \quad \text{for all } v \in V.$

Let $v, w, \in V$ be given. Then

$$w + U = v + U \quad \Rightarrow \quad w \in v + U \quad \Rightarrow \quad (w \in v + U \text{ and } v \in w - U) \Rightarrow$$

$$\Rightarrow \quad (w + U \subset v + U + U \subset v + U$$
$$\text{and } v + U \subset w - U + U \subset w + U) \Rightarrow$$

$$\Rightarrow \quad w + U = v + U,$$

$$w \in \Omega^<(\{\Omega(v)\}) \quad \Leftrightarrow \quad \Omega(w) = \Omega(v) \quad \Leftrightarrow \quad w + U = v + U.$$

Thus $w \in \Omega^<(\{\Omega(v)\})$ if and only if $w \in v + U$, and (14.1) is established.

By (14.1) and *Basic Language*, Section 23 we have

$$\text{Part}\Omega = \{\Omega^<(\{\Omega(v)\}) \mid v \in V\} = \{v + U \mid v \in V\} = V/U.$$

Thus V/U is indeed a partition of V, and (14.1) and *Basic Language*, Section 24 yield

$$\Omega_{V/U}(v) = \Omega_{\text{Part}\Omega}(v) = \Omega^<(\{\Omega(v)\}) = v + U \quad \text{for all } v \in V \text{ (i.e., } \Omega_{V/U} = \Omega). \blacksquare$$

A partition Π of a linear space V is said to be **linear** if $\Pi = V/U$ for some subspace U of V. This choice of terminology will be justified presently (Corollary 14G).

14B. PROPOSITION. *Let the linear space V and the subspaces U and U' of V be given. Then the linear partition V/U is finer than the linear partition V/U' if and only if $U \subset U'$.*

14C. PROPOSITION. *Let the linear mapping L be given.*

(a): $L^<(L_>(A)) = A + \text{Null}L$ *for every subset A of $\text{Dom}L$*

(b): $\mathrm{Part}L = \mathrm{Dom}L/\mathrm{Null}L$.

Proof. For all $v, w \in \mathrm{Dom}L$ we have

$$w \in L^<(\{Lv\}) \quad \Leftrightarrow \quad Lw = Lv \quad \Leftrightarrow \quad L(w-v) = Lw - Lv = 0 \quad \Leftrightarrow$$
$$\Leftrightarrow \quad w - v \in L^<(\{0\}) = \mathrm{Null}L \quad \Leftrightarrow \quad w \in v + \mathrm{Null}L.$$

Therefore

$$L^<(\{Lv\}) = v + \mathrm{Null}L \quad \text{for all } v \in \mathrm{Dom}L.$$

It follows that

$$L^<(L_>(\mathcal{A})) = L^<(\bigcup_{v \in \mathcal{A}} \{Lv\}) = \bigcup_{v \in \mathcal{A}} L^<(\{Lv\}) = \bigcup_{v \in \mathcal{A}}(v + \mathrm{Null}L) = \mathcal{A} + \mathrm{Null}L$$

and

$$\mathrm{Part}L = \{L^<(\{Lv\}) \mid v \in \mathrm{Dom}L\} = \{v + \mathrm{Null}L \mid v \in \mathrm{Dom}L\} = \mathrm{Dom}L/\mathrm{Null}L. \quad \blacksquare$$

14D. COROLLARY. *Let the linear mapping L be given. Then L is injective if and only if* $\mathrm{Null}L = \{0\}$.

Proof. If L is injective, $\mathrm{Null}L = \{0\} + \mathrm{Null}L = L^<(L_>(\{0\}))$ is the singleton $\{0\}$. Conversely, if $\mathrm{Null}L := \{0\}$, then $\mathrm{Part}L = \mathrm{Dom}L/\{0\} = \{v + \{0\} \mid v \in \mathrm{Dom}L\} = \{\{v\} \mid v \in \mathrm{Dom}L\}$, the discrete partition of $\mathrm{Dom}L$, and therefore L is injective. $\blacksquare$

We have shown in Proposition 14C that the partitions of linear mappings are linear partitions. Is every linear partition the partition of some linear mapping? We shall now give an affirmative answer to this question.

14E. LEMMA. *Let the linear space $\mathcal{V}$ and the subspace $\mathcal{U}$ of $\mathcal{V}$ be given. Let $\mathcal{A}, \mathcal{B} \in \mathcal{V}/\mathcal{U}$, $a \in \mathcal{A}, b \in \mathcal{B}$, and $s \in \mathbb{F}$ be given. Then*

$$(14.2) \qquad a + \mathcal{U} = \Omega_{\mathcal{V}/\mathcal{U}}(a) = \mathcal{A} \qquad b + \mathcal{U} = \Omega_{\mathcal{V}/\mathcal{U}}(b) = \mathcal{B}$$

$$(14.3) \quad \mathcal{U} = 0 + \mathcal{U} \in \mathcal{V}/\mathcal{U} \quad \mathcal{A} + \mathcal{B} = (a+b) + \mathcal{U} \in \mathcal{V}/\mathcal{U} \quad -\mathcal{A} = -a + \mathcal{U} \in \mathcal{V}/\mathcal{U}$$

$$(14.4) \qquad\qquad s\mathcal{A} + \mathcal{U} = sa + \mathcal{U} \in \mathcal{V}/\mathcal{U}.$$

Proof. (14.2) follows at once from Proposition 14A. From (14.2) and Remark 13E,(c) we obtain (14.3) and

$$s\mathcal{A} + \mathcal{U} = s(a + \mathcal{U}) + \mathcal{U} = sa + s\mathcal{U} + \mathcal{U} = sa + \mathcal{U} + \mathcal{U} = sa + \mathcal{U} \quad \text{if } s \in \mathbb{F}^\times$$

$$s\mathcal{A} + \mathcal{U} = 0\mathcal{A} + \mathcal{U} = 0 + \mathcal{U} = sa + \mathcal{U} \qquad \text{if } s = 0,$$

so that (14.4) also holds. $\blacksquare$

14F. Theorem. *Let the linear space $\mathcal{V}$ and the subspace $\mathcal{U}$ of $\mathcal{V}$ be given. Then the linear partition $\mathcal{V}/\mathcal{U}$ can be endowed in exactly one way with the structure of a linear space in such a way that $\Omega_{\mathcal{V}/\mathcal{U}}$ becomes linear; namely, by requiring*

$$(14.5) \qquad\qquad 0^{\mathcal{V}/\mathcal{U}} := \mathcal{U}$$

$$(14.6) \qquad\qquad \mathrm{add}^{\mathcal{V}/\mathcal{U}}((\mathcal{A},\mathcal{B})) := \mathcal{A}+\mathcal{B} \quad \text{for all } \mathcal{A},\mathcal{B} \in \mathcal{V}/\mathcal{U}$$

$$(14.7) \qquad\qquad \mathrm{opp}^{\mathcal{V}/\mathcal{U}}(\mathcal{A}) := -\mathcal{A} \quad \text{for all } \mathcal{A} \in \mathcal{V}/\mathcal{U}$$

$$(14.8) \qquad\qquad \mathrm{mult}^{\mathcal{V}/\mathcal{U}}{}_s(\mathcal{A}) := s\mathcal{A}+\mathcal{U} \quad \text{for all } s \in \mathbb{F} \text{ and } \mathcal{A} \in \mathcal{V}/\mathcal{U}.$$

Moreover, we then have $\mathrm{Null}\Omega_{\mathcal{V}/\mathcal{U}} = \mathcal{U}$ *and* $\mathrm{Part}\Omega_{\mathcal{V}/\mathcal{U}} = \mathcal{V}/\mathcal{U}$.

Proof. 1. Let a linear-space structure on $\mathcal{V}/\mathcal{U}$ be given such that $\Omega_{\mathcal{V}/\mathcal{U}}$ is linear. Then $0^{\mathcal{V}/\mathcal{U}} = \Omega_{\mathcal{V}/\mathcal{U}}(0) = 0+\mathcal{U} = \mathcal{U}$, by Proposition 14A, so that (14.5) holds. To prove (14.6), (14.7), (14.8), let $\mathcal{A},\mathcal{B} \in \mathcal{V}/\mathcal{U}$ and $s \in \mathbb{F}$ be given, and choose $a \in \mathcal{A}$, $b \in \mathcal{B}$. Using Lemma 14E and the linearity of $\Omega_{\mathcal{V}/\mathcal{U}}$, we indeed have

$$\mathrm{add}^{\mathcal{V}/\mathcal{U}}((\mathcal{A},\mathcal{B})) = \mathrm{add}^{\mathcal{V}/\mathcal{U}}((\Omega_{\mathcal{V}/\mathcal{U}}(a),\Omega_{\mathcal{V}/\mathcal{U}}(b))) = \Omega_{\mathcal{V}/\mathcal{U}}(a+b) = (a+b)+\mathcal{U} = \mathcal{A}+\mathcal{B}$$

$$\mathrm{opp}^{\mathcal{V}/\mathcal{U}}(\mathcal{A}) = \mathrm{opp}^{\mathcal{V}/\mathcal{U}}(\Omega_{\mathcal{V}/\mathcal{U}}(a)) = \Omega_{\mathcal{V}/\mathcal{U}}(-a) = -a+\mathcal{U} = -\mathcal{A}$$

$$\mathrm{mult}^{\mathcal{V}/\mathcal{U}}{}_s(\mathcal{A}) = \mathrm{mult}^{\mathcal{V}/\mathcal{U}}{}_s(\Omega_{\mathcal{V}/\mathcal{U}}(a)) = \Omega_{\mathcal{V}/\mathcal{U}}(sa) = sa+\mathcal{U} = s\mathcal{A}+\mathcal{U},$$

as was to be shown. We have proved that there is at most one linear-space structure on $\mathcal{V}/\mathcal{U}$ such that $\Omega_{\mathcal{V}/\mathcal{U}}$ is linear, and that this structure, if it exists, must be given by (14.5), (14.6), (14.7), (14.8).

2. We now define $0^{\mathcal{V}/\mathcal{U}}$, $\mathrm{add}^{\mathcal{V}/\mathcal{U}}$, $\mathrm{opp}^{\mathcal{V}/\mathcal{U}}$, $\mathrm{mult}^{\mathcal{V}/\mathcal{U}}$ by (14.5), (14.6), (14.7), (14.8), respectively, as we may by Lemma 14E. We have to show that this prescription satisfies the defining laws for a linear space, and that $\Omega_{\mathcal{V}/\mathcal{U}}$ is linear. The associative and commutative laws for addition are obviously satisfied under (14.6). To prove the neutrality law for addition, the law of opposites, and (LS1)–(LS4) for this prescription, we let $\mathcal{A},\mathcal{B} \in \mathcal{V}/\mathcal{U}$ and $s,t \in \mathbb{F}$ be given, and choose $a \in \mathcal{A}, b \in \mathcal{B}$. From Lemma 14E we then have

$$\mathrm{add}^{\mathcal{V}/\mathcal{U}}((\mathcal{A},0^{\mathcal{V}/\mathcal{U}})) = \mathcal{A}+\mathcal{U} = (a+0)+\mathcal{U} = a+\mathcal{U} = \mathcal{A}$$

$$\mathrm{add}^{\mathcal{V}/\mathcal{U}}((\mathcal{A},\mathrm{opp}^{\mathcal{V}/\mathcal{U}}(\mathcal{A}))) = \mathcal{A}+(-\mathcal{A}) = (a+(-a))+\mathcal{U} = 0+\mathcal{U} = \mathcal{U} = 0^{\mathcal{V}/\mathcal{U}}$$

$$\mathrm{mult}^{\mathcal{V}/\mathcal{U}}{}_s(\mathrm{add}^{\mathcal{V}/\mathcal{U}}((\mathcal{A},\mathcal{B}))) = s(\mathcal{A}+\mathcal{B})+\mathcal{U} = s(a+b)+\mathcal{U} = (sa+sb)+\mathcal{U} =$$

$$= (sa+\mathcal{U})+(sb+\mathcal{U}) = (s\mathcal{A}+\mathcal{U})+(s\mathcal{B}+\mathcal{U}) =$$

$$= \mathrm{add}^{\mathcal{V}/\mathcal{U}}((\mathrm{mult}^{\mathcal{V}/\mathcal{U}}{}_s(\mathcal{A}),\mathrm{mult}^{\mathcal{V}/\mathcal{U}}{}_s(\mathcal{B})))$$

$$\mathrm{mult}^{\mathcal{V}/\mathcal{U}}{}_{s+t}(\mathcal{A}) \quad = (s+t)\mathcal{A} + \mathcal{U} = (s+t)a + \mathcal{U} = (sa+ta) + \mathcal{U} =$$

$$= s(sa+\mathcal{U}) + (ta+\mathcal{U}) = (s\mathcal{A}+\mathcal{U}) + (t\mathcal{A}+\mathcal{U}) =$$

$$= \mathrm{add}^{\mathcal{V}/\mathcal{U}}((\mathrm{mult}^{\mathcal{V}/\mathcal{U}}{}_{s}(\mathcal{A}), \ \mathrm{mult}^{\mathcal{V}/\mathcal{U}}{}_{t}(\mathcal{A}))$$

$$\mathrm{mult}^{\mathcal{V}/\mathcal{U}}{}_{st}(\mathcal{A}) \quad = (st)\mathcal{A} + \mathcal{U} = (st)a + \mathcal{U} = s(ta) + \mathcal{U} =$$

$$= s(t\mathcal{A}+\mathcal{U}) + \mathcal{U} = (\mathrm{mult}^{\mathcal{V}/\mathcal{U}}{}_{s}(\mathrm{mult}^{\mathcal{V}/\mathcal{U}}{}_{t}(\mathcal{A}))$$

(since $ta \in t\mathcal{A} + \mathcal{U} \in \mathcal{V}/\mathcal{U}$), and

$$\mathrm{mult}^{\mathcal{V}/\mathcal{U}}{}_{1}(\mathcal{A}) = 1\mathcal{A} + \mathcal{U} = \mathcal{A} + \mathcal{U} = \mathcal{A}.$$

Thus, the defining laws for a linear space hold. By Lemma 14E we have, for all $v, w \in \mathcal{V}$ and $s \in \mathbb{F}$.

$$\Omega_{\mathcal{V}/\mathcal{U}}(v+w) \quad = (v+w) + \mathcal{U} = (v+\mathcal{U}) + (w+\mathcal{U}) = \Omega_{\mathcal{V}/\mathcal{U}}(v) + \Omega_{\mathcal{V}/\mathcal{U}}(w) =$$

$$= \mathrm{add}^{\mathcal{V}/\mathcal{U}}((\Omega_{\mathcal{V}/\mathcal{U}}(v), \Omega_{\mathcal{V}/\mathcal{U}}(w)))$$

$$\Omega_{\mathcal{V}/\mathcal{U}}(sv) \quad = sv + \mathcal{U} = s(v+\mathcal{U}) + \mathcal{U} = s\Omega_{\mathcal{V}/\mathcal{U}}(v) + \mathcal{U} = \mathrm{mult}^{\mathcal{V}/\mathcal{U}}{}_{s}(\Omega_{\mathcal{V}/\mathcal{U}}(v)),$$

so that $\Omega_{\mathcal{V}/\mathcal{U}}$ is linear.

With this structure in place, $\mathrm{Null}\Omega_{\mathcal{V}/\mathcal{U}} = \Omega_{\mathcal{V}/\mathcal{U}}{}^{<}(\{\mathcal{U}\}) = \mathcal{U}$, and $\mathrm{Part}\Omega_{\mathcal{V}/\mathcal{U}} = \mathcal{V}/\mathcal{U}$ (Proposition 14C, or *Basic Language*, Section 24). ∎

14G. Corollary. *Let the linear space $\mathcal{V}$ be given. Every subspace of $\mathcal{V}$ is the null-space of some surjective linear mapping with domain $\mathcal{V}$. A partition of $\mathcal{V}$ is the partition of a [surjective] linear mapping with domain $\mathcal{V}$ if and only if it is a linear partition of $\mathcal{V}$.*

Proof. Theorem 14F and Proposition 14C. ∎

14H. Remarks. (a): It is clear that (14.8) implies that $\mathrm{mult}^{\mathcal{V}/\mathcal{U}}{}_{s}(\mathcal{A}) = s\mathcal{A}$ for all $s \in \mathbb{F}^{\times}$ and $\mathcal{A} \in \mathcal{V}/\mathcal{U}$, but that $\mathrm{mult}^{\mathcal{V}/\mathcal{U}}{}_{0}(\mathcal{A}) = \mathcal{U}$ for all $\mathcal{A} \in \mathcal{V}/\mathcal{U}$. Thus $0\mathcal{A}$ is not in general $\mathcal{A}$ multiplied by 0 according to the scalar multiplication in the linear space $\mathcal{V}/\mathcal{U}$, in exceptional derogation from the general rules of notation for linear spaces.

(b): Corollary 14G may be compared with the following valid assertion: *Let the linear space $\mathcal{V}$ be given. A subset of $\mathcal{V}$ is the range of a[n injective] linear mapping with codomain $\mathcal{V}$ if and only if it is a subspace of $\mathcal{V}$.* (Theorem 13P for the "only if" part, inclusion mappings for the "if" part.)

(c): If $\mathcal{V}$ is a linear space and $\mathcal{U}$ is a subspace of $\mathcal{V}$, the linear partition $\mathcal{V}/\mathcal{U}$ of $\mathcal{V}$ will always be regarded as endowed with the linear-space structure defined by (14.5)–(14.8). (This linear space is usually called the *quotient-space of $\mathcal{V}$ with respect to* the subspace $\mathcal{U}$, but the term "*linear partition*" seems more suggestive.) ∎

15. Supplements

Let the linear space $\mathcal{V}$ be given. The subspaces $\mathcal{U}$ and $\mathcal{U}'$ of $\mathcal{V}$ are said to be **disjunct**, and $\mathcal{U}'$ is said to be **disjunct from** $\mathcal{U}$, if $\mathcal{U} \cap \mathcal{U}' = \{0\}$. (The term usually encountered is "disjoint"; however, both subspaces contain 0 and cannot therefore be disjoint as sets, so that this usage produces a clash to be avoided.)

The subspaces $\mathcal{U}$ and $\mathcal{U}'$ are said to be **supplementary in** $\mathcal{V}$, and the pair $(\mathcal{U}, \mathcal{U}')$ is also said to be **supplementary in** $\mathcal{V}$, if $\mathcal{U} \cap \mathcal{U}' = \{0\}$ and $\mathcal{U} + \mathcal{U}' = \mathcal{V}$. Given the subspace $\mathcal{U}$ of $\mathcal{V}$, a subspace $\mathcal{U}'$ of $\mathcal{V}$ is said to be **supplementary to** $\mathcal{U}$ **in** $\mathcal{V}$, and is called a **supplement of** $\mathcal{U}$ **in** $\mathcal{V}$, if $\mathcal{U}$ and $\mathcal{U}'$ are supplementary in $\mathcal{V}$. (Terms often encountered are "complementary" and "complement", but this usage clashes with the meaning of "complement" for subsets.)

15A. REMARKS. (a): If $\mathcal{V}$ is a linear space, the subspaces $\mathcal{U}$ and $\mathcal{U}'$ of $\mathcal{V}$ are supplementary in $\mathcal{V}$ if and only if $\mathcal{U}'$ and $\mathcal{U}$ are supplementary in $\mathcal{V}$.

(b): If $\mathcal{V}$ is a linear space, the subspaces $\mathcal{U}$ and $\mathcal{U}'$ of $\mathcal{V}$ are disjunct if and only if they are supplementary in $\mathcal{U} + \mathcal{U}'$. ∎

Let the linear space $\mathcal{V}$ and the subspaces $\mathcal{U}$ and $\mathcal{U}'$ of $\mathcal{V}$ be given. For every $v \in \mathcal{V}$ we examine the following equation

$$?(u, u') \in \mathcal{U} \times \mathcal{U}', \quad v = u + u'.$$

Obviously, this equation has at least one solution for every $v \in \mathcal{V}$ if and only if $\mathcal{U} + \mathcal{U}' = \mathcal{V}$. Our next result determines when this equation has at most one, or exactly one, solution for every $v \in \mathcal{V}$.

15B. PROPOSITION. *Let the linear space $\mathcal{V}$ and the subspaces $\mathcal{U}$ and $\mathcal{U}'$ of $\mathcal{V}$ be given.*

(a): $\mathcal{U}$ and $\mathcal{U}'$ are disjunct if and only if the linear mapping
$((u, u') \mapsto u + u') : \mathcal{U} \times \mathcal{U}' \to \mathcal{V}$ is injective.

(b): $\mathcal{U}$ and $\mathcal{U}'$ are supplementary in $\mathcal{V}$ if and only if the linear mapping
$((u, u') \mapsto u + u') : \mathcal{U} \times \mathcal{U}' \to \mathcal{V}$ is bijective.

Proof. The mapping $S : \mathcal{U} \times \mathcal{U}' \to \mathcal{V}$ defined by the rule

$$(15.1) \qquad S((u, u')) := u + u' \quad \text{for all } (u, u') \in \mathcal{U} \times \mathcal{U}'$$

is the restriction of the linear mapping $\mathrm{add}^{\mathcal{V}} : \mathcal{V} \times \mathcal{V} \to \mathcal{V}$ to the subspace $\mathcal{U} \times \mathcal{U}'$ of $\mathcal{V} \times \mathcal{V}$; it is therefore linear (Example 12G,(f)). For all $(u, u') \in \mathcal{U} \times \mathcal{U}'$ we have

$$S((u, u')) = 0 \quad \Leftrightarrow \quad u + u' = 0 \quad \Leftrightarrow \quad u = -u' \in \mathcal{U} \cap \mathcal{U}'.$$

Therefore $\mathcal{U} \cap \mathcal{U}' = \{0\}$ if and only if $\mathrm{Null}S = \{(0, 0)\}$; by Corollary 14D, this is the case if and only if S is injective. This proves (a). Since $\mathrm{Rng}S = \mathcal{U} + \mathcal{U}'$, (b) follows at once from (a). ∎

15C. PROPOSITION. *Let the linear space* $\mathcal{V}$ *and the subspaces* $\mathcal{U}$ *and* $\mathcal{U}'$ *of* $\mathcal{V}$ *be given.*

(a): *The following statements are equivalent.*

(i): $\mathcal{U}$ *and* $\mathcal{U}'$ *are supplementary in* $\mathcal{V}$.

(ii): *There exists* $P \in \mathrm{Lin}(\mathcal{V}, \mathcal{U})$ *such that* $P|_{\mathcal{U}} = 1_{\mathcal{U}}$ *and* $\mathrm{Null}P = \mathcal{U}'$.

(iii): *There exists* $E \in \mathrm{Lin}\mathcal{V}$ *such that* E *is idempotent (i.e.,* $EE = E$*),* $\mathrm{Fix}E = \mathrm{Rng}E = \mathcal{U}$, *and* $\mathrm{Null}E = \mathcal{U}'$.

(b): *Assume that* $1 + 1 \neq 0$ *in* $\mathbb{F}$. *Then* (i), (ii), (iii) *are equivalent to the following statement.*

(iv): *There exists* $L \in \mathrm{Lin}\mathcal{V}$ *such that* L *is involutory (i.e.,* $LL = 1_{\mathcal{V}}$*) and* $\mathcal{U} = \{v \in \mathcal{V} \mid Lv = v\}$, $\mathcal{U}' = \{v \in \mathcal{V} \mid Lv = -v\}$.

Proof. Proof of (a). (i) *implies* (ii). We define the projection $\pi \in \mathrm{Lin}(\mathcal{U} \times \mathcal{U}', \mathcal{U})$ by the rule $\pi((u, u')) := u$ for all $(u, u') \in \mathcal{U} \times \mathcal{U}'$. Since $\mathcal{U}$ and $\mathcal{U}'$ are supplementary in $\mathcal{V}$, the linear mapping $S \in \mathrm{Lin}(\mathcal{U} \times \mathcal{U}', \mathcal{V})$ defined by (15.1) is bijective (Proposition 15B,(b)) and its inverse S^{-1} is linear (Corollary 12E). The linear mapping $P := \pi S^{-1} \in \mathrm{Lin}(\mathcal{V}, \mathcal{U})$ satisfies $Pu = \pi S^{-1}(u + 0) = \pi((u, 0)) = u$ for all $u \in \mathcal{U}$, so that $P|_{\mathcal{U}} = 1_{\mathcal{U}}$; and $\mathrm{Null}P = (S^{-1})^{<}(\mathrm{Null}\pi) = S_{>}(\{0\} \times \mathcal{U}') = \mathcal{U}'$.

(ii) *implies* (iii). With $P \in \mathrm{Lin}(\mathcal{V}, \mathcal{U})$ as described in (ii), we set $E := P|^{\mathcal{V}} = 1_{\mathcal{U} \subset \mathcal{V}}P \in \mathrm{Lin}\mathcal{V}$. Then $EE = 1_{\mathcal{U} \subset \mathcal{V}}P1_{\mathcal{U} \subset \mathcal{V}}P = 1_{\mathcal{U} \subset \mathcal{V}}1_{\mathcal{U}}P = E$. Now P is right-invertible ($1_{\mathcal{U} \subset \mathcal{V}}$ is a right-inverse), hence surjective, and therefore $\mathrm{Rng}E = \mathrm{Rng}P = \mathcal{U}$. By *Basic Language*, Proposition 26C, we have $\mathrm{Fix}E = \mathrm{Rng}E = \mathcal{U}$. Finally, $\mathrm{Null}E = \mathrm{Null}P = \mathcal{U}'$.

(iii): *implies* (i). Since E is idempotent, $\mathrm{Rng}E$ is the set of fixed points of E. For every $v \in \mathcal{U} \cap \mathcal{U}' = \mathrm{Rng}E \cap \mathrm{Null}E$ we have $v = Ev = 0$; therefore $\mathcal{U} \cap \mathcal{U}' = \{0\}$. For every $v \in \mathcal{V}$ we have $E(v - Ev) = Ev - EEv = Ev - Ev = 0$, and therefore $v - Ev \in \mathrm{Null}E$; consequently $v = Ev + (v - Ev) \in \mathrm{Rng}E + \mathrm{Null}E = \mathcal{U} + \mathcal{U}'$. Therefore $\mathcal{U} + \mathcal{U}' = \mathcal{V}$.

Proof of (b). Assume that $1 + 1 \neq 0$, and set $2 := 1 + 1$.

(iii) *implies* (iv). With $E \in \mathrm{Lin}\mathcal{V}$ as described in (iii), we set $L := 2E - 1_{\mathcal{V}} \in \mathrm{Lin}\mathcal{V}$. Then $LL = 2^2 EE - 2^2 E + 1_{\mathcal{V}} = 1_{\mathcal{V}}$. Let $v \in \mathcal{V}$ be given. Then

$$
\begin{array}{llllllll}
Lv = v & \Leftrightarrow & 2(E - 1_{\mathcal{V}})v = 0 & \Leftrightarrow & Ev = v & \Leftrightarrow & v \in \mathrm{Fix}E = \mathcal{U} \\
Lv = -v & \Leftrightarrow & 2Ev = 0 & \Leftrightarrow & Ev = 0 & \Leftrightarrow & v \in \mathrm{Null}E = \mathcal{U}'.
\end{array}
$$

(iv) *implies* (iii). With $L \in \mathrm{Lin}\mathcal{V}$ as described in (iv), we set $E := \frac{1}{2}(L + 1_{\mathcal{V}}) \in \mathrm{Lin}\mathcal{V}$. Then $EE = (\frac{1}{2})^2 LL + \frac{1}{2}L + (\frac{1}{2})^2 1_{\mathcal{V}} = \frac{1}{2}(L + 1_{\mathcal{V}}) = E$. Let $v \in \mathcal{V}$ be given. Then

$$
\begin{array}{llllllll}
v \in \mathcal{U} & \Leftrightarrow & Lv = v & \Leftrightarrow & Ev = \frac{1}{2}(Lv + v) = v & \Leftrightarrow & v \in \mathrm{Fix}E \\
v \in \mathcal{U}' & \Leftrightarrow & Lv = -v & \Leftrightarrow & Ev = \frac{1}{2}(Lv + v) = 0 & \Leftrightarrow & v \in \mathrm{Null}E. \ \blacksquare
\end{array}
$$

15D. Proposition. *Let the linear space $\mathcal{V}$ and the subspaces $\mathcal{U}$ and $\mathcal{U}'$ of $\mathcal{V}$ be given. Then $\mathcal{U}'$ is a supplement of $\mathcal{U}$ in $\mathcal{V}$ if and only if $\mathcal{U}'$ is maximal among all subspaces of $\mathcal{V}$ disjunct from $\mathcal{U}$, i.e., a maximal member of*

$$(15.2) \qquad\qquad \Sigma := \{\mathcal{W} \in \mathrm{Subsp}(\mathcal{V}) \mid \mathcal{U} \cap \mathcal{W} = \{0\}\}$$

ordered by inclusion.

Proof. *Proof of the "only if" part.* Since $\mathcal{U}'$ is a supplement of $\mathcal{U}$ in $\mathcal{V}$ we have $\mathcal{U}' \in \Sigma$. Let $\mathcal{X} \in \Sigma$ be given, and assume that $\mathcal{U}' \subset \mathcal{X}$; we are to show that $\mathcal{X} \subset \mathcal{U}'$. Let $v \in \mathcal{X}$ be given. Since $\mathcal{U} + \mathcal{U}' = \mathcal{V}$, we may choose $u' \in \mathcal{U}'$ such that $v - u' \in \mathcal{U}$. But $v, u' \in \mathcal{X}$, and hence $v - u' \in \mathcal{U} \cap \mathcal{X} = \{0\}$. Therefore $v = u' \in \mathcal{U}'$. Since $v \in \mathcal{X}$ was arbitrary, we conclude that $\mathcal{X} \subset \mathcal{U}'$, as was to be shown.

Proof of the "if" part. Since $\mathcal{U}' \in \Sigma$, we have $\mathcal{U} \cap \mathcal{U}' = \{0\}$. Let $v \in \mathcal{V}$ be given; we are to show that $v \in \mathcal{U} + \mathcal{U}'$. If $v \in \mathcal{U}' = \{0\} + \mathcal{U}' \subset \mathcal{U} + \mathcal{U}'$, this is valid; we shall therefore assume that $v \in \mathcal{V} \backslash \mathcal{U}'$. We set $\mathcal{X} := \mathcal{U}' + \mathbb{F}v = \mathcal{U}' + \mathrm{Lsp}\{v\} \in \mathrm{Subsp}(\mathcal{V})$ (Proposition 13T). Then $\mathcal{U}' \subsetneqq \mathcal{X}$; since $\mathcal{U}'$ was a maximal member of Σ, we have $\mathcal{X} \notin \Sigma$, so that $\mathcal{U} \cap \mathcal{X} \neq \{0\}$. We may therefore choose $u \in (\mathcal{U} \cap \mathcal{X})^{\times}$. Since $u \in \mathcal{X} = \mathcal{U}' + \mathbb{F}v$, we may choose $s \in \mathbb{F}$ such that $u - sv \in \mathcal{U}'$. Since $u \notin \{0\} = \mathcal{U} \cap \mathcal{U}'$, we cannot have $s = 0$. Therefore $v = \frac{1}{s}(u - (u - sv)) \in \frac{1}{s}(\mathcal{U} + \mathcal{U}') \subset \mathcal{U} + \mathcal{U}'$, as was to be shown. ∎

15E. Proposition. *Let the linear space $\mathcal{V}$ and the subspaces $\mathcal{U}$ and $\mathcal{U}'$ of $\mathcal{V}$ be given. A subspace $\mathcal{W}$ of $\mathcal{U}'$ is a supplement of $\mathcal{U} \cap \mathcal{U}'$ in $\mathcal{U}'$ if and only if $\mathcal{W}$ is a supplement of $\mathcal{U}$ in $\mathcal{U} + \mathcal{U}'$.*

Proof. We have $\mathcal{U}' \cap \mathcal{W} = \mathcal{W}$, and therefore $\mathcal{U} \cap \mathcal{U}' \cap \mathcal{W} = \mathcal{U} \cap \mathcal{W}$, so that $(\mathcal{U} \cap \mathcal{U}') \cap \mathcal{W} = \{0\}$ if and only if $\mathcal{U} \cap \mathcal{W} = \{0\}$.

If $(\mathcal{U} \cap \mathcal{U}') + \mathcal{W} = \mathcal{U}'$, then $\mathcal{U} + \mathcal{U}' = \mathcal{U} + (\mathcal{U} \cap \mathcal{U}') + \mathcal{W} \subset \mathcal{U} + \mathcal{U} + \mathcal{W} \subset \mathcal{U} + \mathcal{W} \subset \mathcal{U} + \mathcal{U}'$, so that $\mathcal{U} + \mathcal{W} = \mathcal{U} + \mathcal{U}'$. Assume conversely, that $\mathcal{U} + \mathcal{W} = \mathcal{U} + \mathcal{U}'$, so that $\mathcal{U}' \subset \mathcal{U} + \mathcal{W}$. Let $v \in \mathcal{U}'$ be given; we may choose $w \in \mathcal{W}$ such that $v - w \in \mathcal{U}$; but $v - w \in \mathcal{U}' - \mathcal{W} \subset \mathcal{U}'$. Therefore $v = (v - w) + w \in (\mathcal{U} \cap \mathcal{U}') + \mathcal{W}$. Since $v \in \mathcal{U}'$ was arbitrary, we find $\mathcal{U}' \subset (\mathcal{U} \cap \mathcal{U}') + \mathcal{W} \subset \mathcal{U}' + \mathcal{U}' \subset \mathcal{U}'$, and so $(\mathcal{U} \cap \mathcal{U}') + \mathcal{W} = \mathcal{U}'$. ∎

We now examine the question whether every subspace of a given linear space has a supplement. We shall see later (Remark 52H) an important class of linear spaces for which an affirmative answer can be given without appealing to the Axiom of Choice or one of its equivalents.

•**15F. Theorem.** *Let the linear space $\mathcal{V}$ and the subspaces $\mathcal{U}$ and $\mathcal{U}'$ of $\mathcal{V}$ be given. There exists a supplement $\mathcal{U}''$ of $\mathcal{U}$ in $\mathcal{V}$ with $\mathcal{U}' \subset \mathcal{U}''$ if (and only if) $\mathcal{U}'$ is disjunct from $\mathcal{U}$.*

Proof. We assume that $\mathcal{U} \cap \mathcal{U}' = \{0\}$ and consider the subcollection Σ of $\mathrm{Subsp}(\mathcal{V})$ defined by (15.2). We regard it as ordered by inclusion and consider the ordered subcollection

$$\Sigma' := \{\mathcal{W} \in \Sigma \mid \mathcal{U}' \subset \mathcal{W}\}.$$

$\Sigma' \neq \emptyset$ since $\mathcal{U}' \in \Sigma'$. Let a non-empty nest $\Gamma \in \mathfrak{P}(\Sigma')$ be given. Then $\bigcup \Gamma \in \mathrm{Subsp}(\mathcal{V})$ by Proposition 13N; we obviously have $\mathcal{U}' \subset \bigcup \Gamma$, and

$$\mathcal{U} \cap \bigcup \Gamma = \bigcup \{\mathcal{U} \cap \mathcal{W} \mid \mathcal{W} \in \Gamma\} = \{0\},$$

so that $\bigcup \Gamma \in \Sigma'$. By a •Set Maximality Principle ((III) in *Basic Language*, Section 172), we may choose a maximal member $\mathcal{U}''$ of Σ'. If $\mathcal{W} \in \Sigma$ satisfies $\mathcal{U}'' \subset \mathcal{W}$, then $\mathcal{U}' \subset \mathcal{U}'' \subset \mathcal{W}$, so that $\mathcal{W} \in \Sigma'$; and since $\mathcal{U}''$ is maximal in Σ', we have $\mathcal{W} = \mathcal{U}''$. Therefore $\mathcal{U}''$ is a maximal member of Σ; by Proposition 15D, $\mathcal{U}''$ is a supplement of $\mathcal{U}$ in $\mathcal{V}$. ■

•15G. COROLLARY. *Let the linear space $\mathcal{V}$ be given. Every subspace of $\mathcal{V}$ has a supplement in $\mathcal{V}$.*

•15H. REMARK. We shall show in •Corollary 21O that the only subspaces of $\mathcal{V}$ with exactly one supplement in $\mathcal{V}$ are $\{0\}$ and $\mathcal{V}$. This could also be proved directly from •Theorem 15F. ■

We conclude this section with a counterpart to Proposition 15D.

•15I. PROPOSITION. *Let the linear space $\mathcal{V}$ and the subspaces $\mathcal{U}$ and $\mathcal{U}'$ of $\mathcal{V}$ be given. Then $\mathcal{U}'$ is a supplement of $\mathcal{U}$ in $\mathcal{V}$ •if and only if $\mathcal{U}'$ is a minimal member of*

$$\Lambda := \{\mathcal{W} \in \text{Subsp}(\mathcal{V}) \mid \mathcal{U} + \mathcal{W} = \mathcal{V}\}$$

ordered by inclusion.

Proof. Proof of the "only if" part. Since $\mathcal{U}'$ is a supplement of $\mathcal{U}$ in $\mathcal{V}$ we have $\mathcal{U}' \in \Lambda$. Let $\mathcal{X} \in \Lambda$ be given, and assume that $\mathcal{X} \subset \mathcal{U}'$; we are to show that $\mathcal{X} = \mathcal{U}'$. Now $\mathcal{U} \cap \mathcal{X} \subset \mathcal{U} \cap \mathcal{U}' = \{0\}$; it follows that $\mathcal{X}$ is a supplement of $\mathcal{U}$ in $\mathcal{V}$. By Proposition 15E, $\mathcal{X}$ is a supplement of $\{0\} = \mathcal{U} \cap \mathcal{U}'$ in $\mathcal{U}'$. Therefore $\mathcal{U}' = \{0\} + \mathcal{X} = \mathcal{X}$, as was to be proved.

•Proof of the "if" part. Since $\mathcal{U}' \in \Lambda$, we have $\mathcal{U} + \mathcal{U}' = \mathcal{V}$. •Choose a supplement $\mathcal{Z}$ of $\mathcal{U} \cap \mathcal{U}'$ in $\mathcal{U}'$, as permitted by •Corollary 15G. By Proposition 15E, $\mathcal{Z}$ is a supplement of $\mathcal{U}$ in $\mathcal{U} + \mathcal{U}' = \mathcal{V}$. Therefore $\mathcal{Z} \in \Lambda$, and $\mathcal{Z} \subset \mathcal{U}'$ by construction. Since $\mathcal{U}'$ is a minimal member of Λ, we have $\mathcal{Z} = \mathcal{U}'$, and hence $\mathcal{U}'$ is a supplement of $\mathcal{U}$ in $\mathcal{V}$. ■

Chapter 2

PROPERTIES OF LINEAR MAPPINGS

21. Linear invertibility

In this chapter we examine systematically properties of invertibility, cancellability, and factorization that linear mappings may have. The ideas, and some of the proofs, will be analogous to those discussed for mappings in general in *Basic Language*, Sections 32, 33, 35, 36. Some of the results involve the partitions of the mappings in question; in view of Section 14, these can be rephrased, for linear mappings, in terms of null-spaces.

We have already noted that the only constant linear mappings are the zero-mappings (Remark 12A,(c)).

21A. PROPOSITION. *Let the linear mappings L and M with $\mathrm{Dom}M = \mathrm{Cod}L$ be given. If M is injective, then $\mathrm{Null}(ML) = \mathrm{Null}L$; if L is surjective, then $\mathrm{Rng}(ML) = \mathrm{Rng}M$.*

Proof. From (13.7) and Corollary 14D, if M is injective then $\mathrm{Null}(ML) = L^<(\mathrm{Null}) = L^<(\{0\}) = \mathrm{Null}L$. The assertion concerning ranges is a special case of *Basic Language*, Proposition 32C.R. ∎

21B. PROPOSITION. *Let the linear mapping $L : \mathcal{V} \to \mathcal{W}$ be given. Then there are linear mappings $M : \mathcal{V} \to \mathcal{U}$ and $N : \mathcal{U} \to \mathcal{W}$ such that $L = NM$ and M is surjective and N is injective.*

Proof. Set $\mathcal{U} := \mathrm{Rng}L$, $M := L|^{\mathrm{Rng}}$, $N := 1_{\mathrm{Rng}L \subset \mathcal{W}}$. ∎

21C. PROPOSITION. *Let the linear mapping L and the subspace $\mathcal{U}$ of $\mathrm{Dom}L$ be given. The linear mapping $L|_{\mathcal{U}}^{\mathrm{Rng}L}$ is injective if and only if $\mathcal{U} \cap \mathrm{Null}L = \{0\}$, surjective if and only if $\mathcal{U} + \mathrm{Null}L = \mathrm{Dom}L$, bijective if and only if $\mathcal{U}$ is a supplement of $\mathrm{Null}L$ in $\mathrm{Dom}L$.*

Proof. We have $\mathrm{Null}(L|_{\mathcal{U}}^{\mathrm{Rng}L}) = \mathrm{Null}(L|_{\mathcal{U}}) = \mathcal{U} \cap \mathrm{Null}L$. By Corollary 14D, $L|_{\mathcal{U}}^{\mathrm{Rng}L}$ is injective if and only if $\mathcal{U} \cap \mathrm{Null}L = \{0\}$.

If $\mathcal{U} + \mathrm{Null}L = \mathrm{Dom}L$, Lemma 12H yields

$\mathrm{Rng}L = L_>(\mathcal{U} + \mathrm{Null}L) = L_>(\mathcal{U}) + L_>(\mathrm{Null}L) = L_>(\mathcal{U}) + \{0\} = \mathrm{Rng}(L|_{\mathcal{U}}^{\mathrm{Rng}L})$, so that $L_{\mathcal{U}}^{\mathrm{Rng}L}$ is surjective. Conversely, if $L|_{\mathcal{U}}^{\mathrm{Rng}L}$ is surjective, Proposition 14C,(a) yields $\mathrm{Dom}L = L^<(\mathrm{Rng}L) = L^<(\mathrm{Rng}(L|_{\mathcal{U}}^{\mathrm{Rng}L})) = L^<(L_>(\mathcal{U})) = \mathcal{U} + \mathrm{Null}L.$ ∎

21D. Corollary. *Let the linear space $\mathcal{V}$ and the subspaces $\mathcal{U}$ and $\mathcal{U}'$ of $\mathcal{V}$ be given. Then $\Omega_{\mathcal{V}/\mathcal{U}}|_{\mathcal{U}'} \in \mathrm{Lin}(\mathcal{U}', \mathcal{V}/\mathcal{U})$ is bijective if and only if $\mathcal{U}$ and $\mathcal{U}'$ are supplementary in $\mathcal{V}$.*

Proof. The partition mapping $\Omega_{\mathcal{V}/\mathcal{U}} \in \mathrm{Lin}(\mathcal{V}, \mathcal{V}/\mathcal{U})$ is surjective, and $\mathrm{Null}\Omega_{\mathcal{V}/\mathcal{U}} = \mathcal{U}$ (Theorem 14F). The conclusion follows from Proposition 21C. ∎

21E. Corollary. *Let the linear space $\mathcal{W}$ and the linear mappings L and M with $\mathrm{Dom}M = \mathcal{W} = \mathrm{Cod}L$ be given.*

(a): *ML is injective if and only if L is injective and $\mathrm{Rng}L \cap \mathrm{Null}M = \{0\}$.*

(b): *ML is surjective if and only if M is surjective and $\mathrm{Rng}L + \mathrm{Null}M = \mathcal{W}$.*

(c): *ML is bijective if and only if L is injective, M is surjective, and $\mathrm{Rng}L$ and $\mathrm{Null}M$ are supplementary in $\mathcal{W}$.*

Proof. We have

$$(21.1) \qquad ML = 1_{\mathrm{Rng}M \subset \mathrm{Cod}M}(M|_{\mathrm{Rng}L}^{\mathrm{Rng}M})(L|^{\mathrm{Rng}}).$$

If ML is injective, then $L|^{\mathrm{Rng}}$ is injective, hence bijective; consequently $M|_{\mathrm{Rng}L}^{\mathrm{Rng}M}$ is injective; by Proposition 21C, $\mathrm{Rng}L \cap \mathrm{Null}M = \{0\}$. Conversely, if L is injective and $\mathrm{Rng}L \cap \mathrm{Null}M = \{0\}$, then all three linear mappings in the right-hand side of (21.1) are injective, and hence so is ML.

If ML is surjective, then M is surjective; thus $1_{\mathrm{Rng}M \subset \mathrm{Cod}M} = 1_{\mathrm{Rng}M}$, and consequently $M|_{\mathrm{Rng}L}^{\mathrm{Rng}M}$ is surjective. By Proposition 21C, $\mathrm{Rng}L + \mathrm{Null}M = \mathcal{W}$. Conversely, if M is surjective and $\mathrm{Rng}L + \mathrm{Null}M = \mathcal{W}$, all three linear mappings in the right-hand side of (21.1) are surjective, and hence so is ML. ∎

A linear mapping L is said to be **linearly left-invertible** if there is a linear left-inverse of L, **linearly right-invertible** if there is a linear right-inverse of L, and **linearly invertible** if there is a linear inverse of L.

21F. Remark. By Corollary 12E, a linear mapping is linearly invertible if and only if it is bijective; hence if and only if it is invertible. The term "linearly invertible" is therefore always replaced, without ambiguity, by "invertible".

A linear mapping is also called a **linear isomorphism** if it is invertible. If $\mathcal{V}$ and $\mathcal{W}$ are linear spaces, $\mathcal{V}$ is said to be **(linearly) isomorphic to** $\mathcal{W}$ if there exists a linear isomorphism from $\mathcal{V}$ to $\mathcal{W}$. Since identity mappings of linear spaces and composites of linear isomorphisms are all linear isomorphisms, "is isomorphic to" becomes an equivalence relation in every collection of linear spaces.

If $\mathcal{V}$ and $\mathcal{W}$ are linear spaces, we set

$$\mathrm{Lis}(\mathcal{V}, \mathcal{W}) := \{L \in \mathrm{Lin}(\mathcal{V}, \mathcal{W}) \mid L \text{ is invertible}\},$$

the set of all linear isomorphisms from $\mathcal{V}$ to $\mathcal{W}$. For each linear space $\mathcal{V}$ we set $\mathrm{Lis}\mathcal{V} := \mathrm{Lis}(\mathcal{V}, \mathcal{V})$. The members of $\mathrm{Lis}\mathcal{V}$ are called **linear automorphisms of $\mathcal{V}$.** ∎

21G. Lemma. *Let the linear mapping L be given.*

(a): *If L is injective and $\mathcal{U}$ is a supplement of $\mathrm{Rng}\,L$ in $\mathrm{Cod}\,L$, there is exactly one linear left-inverse M of L such that $\mathrm{Null}\,M = \mathcal{U}$.*

(b): *If L is surjective and $\mathcal{U}$ is a supplement of $\mathrm{Null}\,L$ in $\mathrm{Dom}\,L$, there is exactly one linear right-inverse M of L such that $\mathrm{Rng}\,M = \mathcal{U}$.*

Proof. Proof of (a). By Proposition 15C we may choose $P \in \mathrm{Lin}(\mathrm{Cod}\,L, \mathrm{Rng}\,L)$ such that $P|_{\mathrm{Rng}L} = 1_{\mathrm{Rng}L}$ and $\mathrm{Null}\,P = \mathcal{U}$.

If M is a linear left-inverse of L, we have $1_{\mathrm{Dom}L} = ML = (M|_{\mathrm{Rng}L})(L|^{\mathrm{Rng}})$, so that $(M|_{\mathrm{Rng}L}) = (L|^{\mathrm{Rng}})^{-1}$. If also $\mathrm{Null}\,M = \mathcal{U}$ we have, for every $v \in \mathrm{Cod}\,L$, $P(v - Pv) = Pv - (P|_{\mathrm{Rng}L})(Pv) = Pv - Pv = 0$, and hence $v - Pv \in \mathrm{Null}\,P = \mathcal{U} = \mathrm{Null}\,M$; it follows that

$$Mv = M(v - Pv) + MPv = 0 + (M|_{\mathrm{Rng}L})Pv = (L|^{\mathrm{Rng}})^{-1}Pv.$$

Therefore we must have

$$(21.2) \qquad M = (L|^{\mathrm{Rng}})^{-1}P.$$

Thus there is at most one linear left-inverse of L with null-space $\mathcal{U}$.

Now *define* $M \in \mathrm{Lin}(\mathrm{Cod}\,L, \mathrm{Dom}\,L)$ by (21.2). Then

$$ML = (L|^{\mathrm{Rng}})^{-1}PL = (L|^{\mathrm{Rng}})^{-1}(P|_{\mathrm{Rng}L})(L|^{\mathrm{Rng}}) = (L|^{\mathrm{Rng}})^{-1}1_{\mathrm{Rng}L}(L|^{\mathrm{Rng}}) = 1_{\mathrm{Dom}L},$$

so that M is indeed a linear left-inverse of L. Since $(L|^{\mathrm{Rng}})^{-1}$ is invertible, we also have, from (21.2) that $\mathrm{Null}\,M = \mathrm{Null}\,P = \mathcal{U}$.

Proof of (b). By Proposition 21C, $L|_{\mathcal{U}}$ is invertible. If M is a linear right-inverse of L and $\mathrm{Rng}\,M = \mathcal{U}$, then $1_{\mathrm{Cod}L} = LM = (L|_{\mathrm{Rng}M})(M|^{\mathrm{Rng}}) = (L|_{\mathcal{U}})(M|^{\mathrm{Rng}})$, and therefore we must have $M|^{\mathrm{Rng}} = (L|_{\mathcal{U}})^{-1}$, whence

$$(21.3) \qquad M = (L|_{\mathcal{U}})^{-1}|^{\mathrm{Dom}L}.$$

Thus there is at most one linear right-inverse of L with range $\mathcal{U}$.

Now *define* $M \in \mathrm{Lin}(\mathrm{Cod}\,L, \mathrm{Dom}\,L)$ by (21.3). Then

$$LM = L((L|_{\mathcal{U}})^{-1}|^{\mathrm{Dom}L}) = (L|_{\mathcal{U}})(L|_{\mathcal{U}})^{-1} = 1_{\mathrm{Cod}L},$$

so that M is indeed a linear right-inverse of L. From (21.3) we also have $\mathrm{Rng}\,M = \mathrm{Rng}(L|_{\mathcal{U}})^{-1} = \mathrm{Dom}(L|_{\mathcal{U}}) = \mathcal{U}$. ∎

21H. Theorem. *Let the linear mapping L be given.*

(a): *A subspace of $\mathrm{Cod}\,L$ is the null-space of one (indeed, of exactly one) linear left-inverse of L if and only if it is a supplement of $\mathrm{Rng}\,L$ in $\mathrm{Cod}\,L$ and L is injective.*

(b): *A subspace of $\mathrm{Dom}\,L$ is the range of one (indeed, of exactly one) linear right-inverse of L if and only if it is a supplement of $\mathrm{Null}\,L$ in $\mathrm{Dom}\,L$ and L is surjective.*

Proof. The "if" parts follow from Lemma 21G; the "only if" parts follow from Corollary 21E,(c). ∎

21I. COROLLARY. *Let the linear mapping L be given.*

(a): *L is linearly left-invertible if and only if L is injective and* Rng$L$ *has a supplement in* Cod$L$.

(b): *L is linearly right-invertible if and only if L is surjective and* Null$L$ *has a supplement in* Dom$L$.

21J. EXAMPLE*. With Cont$(\mathbb{R},\mathbb{R})$ and Cont$^1(\mathbb{R},\mathbb{R})$ defined as in Examples 12G,(h) and 13G,(f), we define the linear mappings $D:$ Cont$^1(\mathbb{R},\mathbb{R}) \to$ Cont$(\mathbb{R},\mathbb{R})$ and $J:$ Cont$(\mathbb{R},\mathbb{R}) \to$ Map$(\mathbb{R},\mathbb{R})$ by the rules

$$Df := f^{\cdot} \quad \text{for all } f \in \text{Cont}^1(\mathbb{R},\mathbb{R})$$

$$(Jf)(t) := \int_0^t f \quad \text{for all } t \in \mathbb{R} \text{ and } f \in \text{Cont}(\mathbb{R}.\mathbb{R}).$$

The Fundamental Theorem of the Calculus asserts that Rng$J \subset$ Cont$^1(\mathbb{R},\mathbb{R})$ and that $(Jf)^{\cdot} = f$ for all $f \in$ Cont$(\mathbb{R},\mathbb{R})$, i.e., that $J|^{\text{Cont}^1(\mathbb{R},\mathbb{R})}$ is a linear right-inverse of D. However, D is not linearly left-invertible, since it is not injective. Null$D$ is the subspace $\mathbb{R}(1_{\mathbb{R}\to\mathbb{R}})$ of Cont$^1(\mathbb{R},\mathbb{R})$ consisting of all the constants. In illustration of Corollary 21E,(c), this subspace is a supplement of Rng$J = \{f \in$ Cont$^1(\mathbb{R},\mathbb{R}) \mid f(0) = 0\} =$ Null ev$_0^{\text{Cont}^1(\mathbb{R},\mathbb{R})}$ in Cont$^1(\mathbb{R},\mathbb{R})$. ∎

●**21K. COROLLARY.** *Let the linear mapping L be given.*

(a): *L is linearly left-invertible ●if and only if L is injective.*

(b): *L is linearly right-invertible ●if and only if L is surjective.*

Proof. Corollaries 21I and ●15G. ∎

●**21L. COROLLARY.** *Let the linear spaces $\mathcal{V}$ and $\mathcal{W}$, and the subspace $\mathcal{U}$ of $\mathcal{V}$ be given. For every $M \in$ Lin$(\mathcal{U},\mathcal{W})$ there exists $L \in$ Lin$(\mathcal{V},\mathcal{W})$ such that $L|_{\mathcal{U}} = M$; i.e., the linear mapping $(L \to L|_{\mathcal{U}}):$ Lin$(\mathcal{V},\mathcal{W}) \to$ Lin$(\mathcal{U},\mathcal{W})$ is surjective.*

Proof. $1_{\mathcal{U}\subset\mathcal{V}}$ is injective. By ●Corollary 21K,(a), we may choose a linear left-inverse $P \in$ Lin$(\mathcal{V},\mathcal{U})$ of $1_{\mathcal{U}\subset\mathcal{V}}$. Then $L := MP$ verifies the assertion, since $L|_{\mathcal{U}} = MP1_{\mathcal{U}\subset\mathcal{V}} = M$. ∎

●**21M. PROPOSITION.** *Let the linear spaces $\mathcal{V}$ and $\mathcal{W}$ be given. For every $v \in \mathcal{V}^{\times}$ and every $w \in \mathcal{W}$ there exists $L \in$ Lin$(\mathcal{V},\mathcal{W})$ such that $Lv = w$. Therefore* Lin$(\mathcal{V},\mathcal{W})$ *is a zero-space if and only if $\mathcal{V}$ or $\mathcal{W}$ is a zero-space.*

Proof. Let $v \in \mathcal{V}^{\times}$ and $w \in \mathcal{W}$ be given. Then $v\otimes \in$ Lin$(\mathbb{F},\mathcal{V})$ (Example 12G,(g)) is injective. By ●Corollary 21K,(a) we may choose a linear left-inverse $K \in$ Lin$(\mathcal{V},\mathbb{F})$ of $v\otimes$, and find $Kv = K(v \otimes 1) = 1$. Then $L := w \otimes K \in$ Lin$(\mathcal{V},\mathcal{W})$ satisfies $Lv = w \otimes Kv = w \otimes 1 = w$. The "only if" part of the second assertion follows; the "if" part is trivial. ∎

▼   **21N. PROPOSITION.** *Let the linear space $\mathcal{V}$ and the subspace $\mathcal{U}$ of $\mathcal{V}$ be given. Then $\mathcal{U}$ has exactly one supplement in $\mathcal{V}$ if and only if $\mathcal{U}$ has some supplement in $\mathcal{V}$ and* Lin$(\mathcal{V}/\mathcal{U},\mathcal{U})$ *is a zero-space.*

Proof. The linear mapping $\Omega_{\mathcal{V}/\mathcal{U}} : \mathcal{V} \to \mathcal{V}/\mathcal{U}$ is surjective and $\mathrm{Null}\Omega_{\mathcal{V}/\mathcal{U}} = \mathcal{U}$ (Theorem 14F). By Theorem 21H,(b), the supplements of $\mathcal{U}$ in $\mathcal{V}$ are precisely the ranges of linear right-inverses of $\Omega_{\mathcal{V}/\mathcal{U}}$, and $\mathcal{U}$ has exactly one supplement in $\mathcal{V}$ if and only if $\Omega_{\mathcal{V}/\mathcal{U}}$ has exactly one linear right-inverse. We may therefore stipulate that $\Omega_{\mathcal{V}/\mathcal{U}}$ is linearly right-invertible, and choose a linear right-inverse $K \in \mathrm{Lin}(\mathcal{V}/\mathcal{U}, \mathcal{V})$ of $\Omega_{\mathcal{V}/\mathcal{U}}$. A linear mapping $L \in \mathrm{Lin}(\mathcal{V}/\mathcal{U}, \mathcal{V})$ is a linear-right inverse of $\Omega_{\mathcal{V}/\mathcal{U}}$ if and only if $\Omega_{\mathcal{V}/\mathcal{U}}(L - K) = 0$, hence if and only if $\mathrm{Rng}(L - K) \subset \mathrm{Null}\Omega_{\mathcal{V}/\mathcal{U}} = \mathcal{U}$, hence if and only if $L - K = M|^{\mathcal{V}}$ for some $M \in \mathrm{Lin}(\mathcal{V}/\mathcal{U}, \mathcal{U})$. We conclude that K is the only linear right-inverse of $\Omega_{\mathcal{V}/\mathcal{U}}$ (equivalently, $\mathrm{Rng}K$ is the only supplement of $\mathcal{U}$ in $\mathcal{V}$) if and only if $\mathrm{Lin}(\mathcal{V}/\mathcal{U}, \mathcal{U})$ is a zero-space. ∎

•**21O. Corollary.** *Let the linear space $\mathcal{V}$ and the subspace $\mathcal{U}$ be given. Then $\mathcal{U}$ has exactly one supplement in $\mathcal{V}$ if and only if $\mathcal{U} = \{0\}$ or $\mathcal{U} = \mathcal{V}$.*

Proof. It is trivial that $\mathcal{V}$ is the only supplement of $\{0\}$ in $\mathcal{V}$ and that $\{0\}$ is the only supplement of $\mathcal{V}$ in $\mathcal{V}$. Assume, conversely, that $\mathcal{U}$ has exactly one supplement in $\mathcal{V}$. By Proposition 21N, it follows that $\mathrm{Lin}(\mathcal{V}/\mathcal{U}, \mathcal{U})$ is a zero-space. By •Proposition 21M, either $\mathcal{V}/\mathcal{U}$ is a zero-space, and then $\mathcal{U} = \mathcal{V}$, or $\mathcal{U}$ is a zero-space, and then $\mathcal{U} = \{0\}$. ∎

•**21P. Corollary.** *Let the linear mapping L be given. The following statements are equivalent:*

(i): *L is (linearly) invertible.*

(ii): *L has exactly one linear left-inverse, and if $\mathrm{Dom}L = \{0\}$ then $\mathrm{Cod}L = \{0\}$.*

(iii): *L has exactly one linear right-inverse, and if $\mathrm{Cod}L = \{0\}$ then $\mathrm{Dom}L = \{0\}$.*

Proof. (i) *implies* (ii) *and* (iii). If L is invertible, its inverse is linear, and is the only left-inverse and the only right-inverse of L (*Basic Language*, Proposition 33B). Since L is bijective, $\mathrm{Dom}L = \{0\}$ if and only if $\mathrm{Cod}L = \{0\}$.

•(ii) *implies* (i). By Theorem 21H,(a), L is injective and $\mathrm{Rng}L$ has exactly one supplement in $\mathrm{Cod}L$. By •Corollary 21O, $\mathrm{Rng}L = \{0\}$ or $\mathrm{Rng}L = \mathrm{Cod}L$. In the former case, since K was injective we have $\mathrm{Dom}L = \{0\}$ and hence, by assumption, $\mathrm{Cod}L = \{0\}$. Then $\mathrm{Rng}L = \mathrm{Cod}L$ in either case, and L is surjective. Thus L is bijective, hence invertible.

•(iii) *implies* (i). By Theorem 21H,(b), L is surjective and $\mathrm{Null}L$ has exactly one supplement in $\mathrm{Dom}L$. By •Corollary 21O, $\mathrm{Null}L = \{0\}$ or $\mathrm{Null}L = \mathrm{Dom}L$. In the latter case, $L = 0$; since L was surjective, this implies $\mathrm{Cod}L = \{0\}$, and hence, by assumption, $\mathrm{Dom}L = \{0\}$. Thus $\mathrm{Null}L = \{0\}$ in either case. By Corollary 14D, L is injective. Thus L is bijective, hence invertible. ∎

If the linear space $\mathcal{V}$ is not a zero-space, then $0_{\{0\}\to\mathcal{V}}$ has exactly one (linear) left-inverse, namely $0_{\mathcal{V}\to\{0\}}$, and $0_{\mathcal{V}\to\{0\}}$ has exactly one *linear* right-inverse, namely $0_{\{0\}\to\mathcal{V}}$, (although there are other, non-linear, right-inverses); but neither mapping is invertible.

22. Cancellability and factorization

A linear mapping $L : \mathcal{V} \to \mathcal{W}$ is called a **linear monomorphism** if it is left-cancellable from composition with *linear* mappings; i.e., if for every linear space $\mathcal{U}$ we have

(22.1) $$\forall M, N \in \mathrm{Lin}(\mathcal{U}, \mathcal{V}), \quad LM = LN \quad \Rightarrow \quad M = N;$$

equivalently, by Remark 13I, if for every linear space $\mathcal{U}$ we have

(22.2) $$\forall M \in \mathrm{Lin}(\mathcal{U}, \mathcal{V}), \quad LM = 0 \quad \Rightarrow \quad M = 0.$$

A linear mapping $L : \mathcal{V} \to \mathcal{W}$ is called a **linear epimorphism** if it is right-cancellable from composition with *linear* mappings; i.e., if for every linear space $\mathcal{U}$ we have

(22.3) $$\forall M, N \in \mathrm{Lin}(\mathcal{W}, \mathcal{U}), \quad ML = NL \quad \Rightarrow \quad M = N;$$

equivalently, by Remark 13I, if for every linear space $\mathcal{U}$ we have

(22.4) $$\forall M \in \mathrm{Lin}(\mathcal{W}, \mathcal{U}), \quad ML = 0 \quad \Rightarrow \quad M = 0.$$

We recall that for every linear mapping $L : \mathcal{V} \to \mathcal{W}$ and linear space $\mathcal{U}$ we have

(22.5) $$\mathrm{Rng}(LM) \subset \mathrm{Rng}L \quad \text{for all } M \in \mathrm{Lin}(\mathcal{U}, \mathcal{V})$$

(22.6) $$\mathrm{Null}(ML) \supset \mathrm{Null}L \quad \text{for all } M \in \mathrm{Lin}(\mathcal{W}, \mathcal{U}).$$

A linear mapping $L : \mathcal{V} \to \mathcal{W}$ is called a **linear-embedding** if for every linear mapping $N : \mathcal{U} \to \mathcal{W}$ with $\mathrm{Rng}N \subset \mathrm{Rng}L$ there is exactly one linear mapping $M : \mathcal{U} \to \mathcal{V}$ such that $N = LM$. A linear mapping $L : \mathcal{V} \to \mathcal{W}$ is called a **linear-quotient-mapping** if for every linear mapping $N : \mathcal{V} \to \mathcal{U}$ with $\mathrm{Null}N \supset \mathrm{Null}L$ there is exactly one linear mapping $M : \mathcal{W} \to \mathcal{U}$ such that $N = ML$.

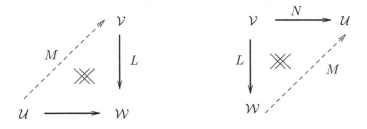

22A.L. Proposition. *If $L : \mathcal{V} \to \mathcal{W}$ and $L' : \mathcal{V}' \to \mathcal{W}$ are linear-embeddings with $\mathrm{Rng}L = \mathrm{Rng}L'$, then the unique linear mappings $M : \mathcal{V}' \to \mathcal{V}$ and $M' : \mathcal{V} \to \mathcal{V}'$*

that satisfy $LM = L'$ *and* $L'M' = L$ *are invertible, and each is the inverse of the other.*

Proof. This is analogous to the proof of *Basic Language*, Proposition 36A.L. ∎

22A.R. PROPOSITION. *If* $L : V \to W$ *and* $L' : V \to W'$ *are linear-quotient-mappings with* $\mathrm{Null}L = \mathrm{Null}L'$, *then the unique linear mappings* $M : W \to W'$ *and* $M' : W' \to W$ *that satisfy* $ML = L'$ *and* $M'L' = L$ *are invertible, and each is the inverse of the other.*

Proof. This is analogous to the proof of Proposition 22A.L. ∎

22B.L. PROPOSITION. *Let the linear mapping* $L : V \to W$ *be given. The following statements are equivalent.*

   (i): *L is a linear-embedding.*

   (ii): *L is a linear monomorphism.*

   (iii): *L is injective.*

Proof. (i) *implies* (ii). Let the linear space $\mathcal{U}$ be given, and let $M, N \in \mathrm{Lin}(\mathcal{U}, V)$ satisfy $LM = LN$. Then $\mathrm{Rng}(LM) = \mathrm{Rng}(LN) \subset \mathrm{Rng}L$. Since L is a linear-embedding, we must have $M = N$.

(ii) *implies* (iii). We have $L1_{\mathrm{Null}L \subset V} = L|_{\mathrm{Null}L} = 0_{\mathrm{Null}L \to W}$. Since L is a linear monomorphism, $1_{\mathrm{Null}L \subset V} = 0_{\mathrm{Null}L \to V}$. Therefore $\mathrm{Null}L = \{0\}$; by Corollary 14D it follows that L is injective.

(iii) *implies* (i). Since L is injective, it is a *set-embedding* (*Basic Language*, Proposition 36B.L.). Let a linear mapping $N : \mathcal{U} \to W$ be given and assume that $\mathrm{Rng}N \subset \mathrm{Rng}L$. Then there exists exactly one mapping $M : \mathcal{U} \to V$ (we do not yet know whether M is linear) such that $N = LM$. By Proposition 12D,(b), however, it follows that M is in fact linear. ∎

22B.R. PROPOSITION. *Let the linear mapping* $L : V \to W$ *be given. The following statements are equivalent:*

   (i): *L is a linear-quotient-mapping;*

   (ii): *L is a linear epimorphism;*

   (iii): *L is surjective.*

Proof. (i) *implies* (ii). Let the linear space $\mathcal{U}$ be given, and let $M, N \in \mathrm{Lin}(W, \mathcal{U})$ satisfy $ML = NL$. Then $\mathrm{Null}(ML) = \mathrm{Null}(NL) \supset \mathrm{Null}L$. Since L is a linear-quotient-mapping, we must have $M = N$.

(ii) *implies* (iii). We consider the linear partition $W/\mathrm{Rng}L$. We have $\mathrm{Rng}(\Omega_{W/\mathrm{Rng}L}L) = (\Omega_{W/\mathrm{Rng}L})_{>}(\mathrm{Rng}L) = \{\mathrm{Rng}L\} = \{0^{W/\mathrm{Rng}L}\}$, by Theorem 14F. Therefore $\Omega_{W/\mathrm{Rng}L}L = 0_{V \to W/\mathrm{Rng}L}$. Since L is a linear epimorphism, we have $\Omega_{W/\mathrm{Rng}L} = 0_{W \to W/\mathrm{Rng}L}$, which means that $\mathrm{Rng}L = \mathrm{Null}\Omega_{W/\mathrm{Rng}L} = W$. Therefore L is surjective.

(iii) *implies* (i). Since L is surjective, it is a *set-quotient-mapping* (*Basic Language*, Proposition 36B.R). Let a linear mapping $N : V \to \mathcal{U}$ be given and assume that $\mathrm{Null}N \supset \mathrm{Null}L$. By Proposition 14B we have $\mathrm{Part}N \sqsubset \mathrm{Part}L$. Then there exists exactly one mapping $M : W \to \mathcal{U}$ (we do not yet know whether M is linear) such that $N = M \circ L$. By Proposition 12D,(a), however, it follows that M is in fact linear. ■

22C. REMARKS. (a): Propositions 22B.L and 22B.R in conjunction with *Basic Language*, Propositions 36B.L and 36B.R, show that for given linear spaces $\mathcal{V}$ and $\mathcal{W}$ the linear monomorphisms, linear epimorphisms, linear-embeddings, linear-quotient-mappings from $\mathcal{V}$ to $\mathcal{W}$ are precisely the set-monomorphisms (left-cancellable mappings), set-epimorphisms (right-cancellable mappings), (set-)embeddings, (set-)quotient-mappings, respectively, from $\mathcal{V}$ to $\mathcal{W}$ that happen to be linear. We may, in particular, omit the hyphen in "**linear embedding**" and "**linear quotient-mapping**", without ambiguity.

(b): The term "*(linear) quotient-mapping*" is often found attached exclusively to the (linear) partition mappings of linear partitions (cf. Section 14). Inasmuch as these are linear quotient-mappings – as defined here – obtained by a standard procedure from prescribed domains and null spaces (cf. Corollary 14G), they may be called **standard linear quotient-mappings**. By analogy, the (linear) inclusion mappings of subspaces, which are linear embeddings obtained by a standard procedure from prescribed codomains and ranges, may be called **standard linear embeddings**. ■

22D. THEOREM. *Let the linear mappings $M : \mathcal{V} \to \mathcal{V}'$ and $N : \mathcal{W}' \to \mathcal{W}$ be given, and assume that M is surjective and N is injective.*

(a): *For a given $L \in \mathrm{Lin}(\mathcal{V}, \mathcal{W})$ there is at most one $L' \in \mathrm{Lin}(\mathcal{V}', \mathcal{W}')$ such that $L = NL'M$; such an L' exists if and only if* $\mathrm{Null}L \supset \mathrm{Null}M$ *and* $\mathrm{Rng}L \subset \mathrm{Rng}N$.

(b): *This linear mapping L' is injective if and only if* $\mathrm{Null}L = \mathrm{Null}M$*, and is surjective if and only if* $\mathrm{Rng}L = \mathrm{Rng}N$.

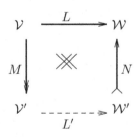

Proof. This is entirely analogous to the proof of *Basic Language*, Theorem 36C. We use (22.5), (22.6), and Propositions 21A, 21B, 22B.L, and 22B.R. ■

22E. COROLLARY. *Let the linear mapping L be given. Then there is exactly one linear mapping $L' : \mathrm{Dom}L/\mathrm{Null}L \to \mathrm{Rng}L$ such that $L = 1_{\mathrm{Rng}L\subset\mathrm{Cod}L}L'\Omega_{\mathrm{Dom}L/\mathrm{Null}L}$; this linear mapping L' is invertible.* ■

▼ Inspired by the discussion in *Basic Language*, Section 35, we take a closer look at the notion of cancellability from composition with linear mappings. A linear mapping $L : \mathcal{V} \to \mathcal{W}$ is said to be **linearly left-cancellable with respect to** the linear space $\mathcal{U}$ if (22.1) or, equivalently, (22.2) holds; L is said to be **linearly right-cancellable with respect to** $\mathcal{U}$ if (22.3) or, equivalently, (22.4) holds. Thus L is a linear monomorphism [linear epimorphism] if and only if L is linearly left-[right-] cancellable with respect to *every* linear space $\mathcal{U}$.

We note that every linear mapping $L : \mathcal{V} \to \mathcal{W}$ is both linearly left-cancellable and linearly right-cancellable with respect to a zero-space $\{0\}$, since both $\mathrm{Lin}(\{0\}, \mathcal{V}) = \{0_{\{0\} \to \mathcal{V}}\}$ and $\mathrm{Lin}(\mathcal{W}, \{0\}) = \{0_{\mathcal{W} \to \{0\}}\}$ are singletons.

•22F.L. PROPOSITION. *Let the linear mapping L be given. The following statements are equivalent:*

(i): *L is injective.*

(ii): *L is a linear monomorphism.*

(iii): *L is linearly left-cancellable with respect to $\mathbb{F}$.*

(iv): *L is linearly left-cancellable with respect to some non-trivial linear space.*

Proof. The implication (i) $\Rightarrow$ (ii) is part of Proposition 17B.L. The implications (ii) $\Rightarrow$ (iii) $\Rightarrow$ (iv) are trivial. We shall prove (iii) $\Rightarrow$ (i) and •(iv) $\Rightarrow$ (iii).

(iii) *implies* (i). Let $v \in \mathrm{Null}L$ be given. Then $v\otimes \in \mathrm{Lin}(\mathbb{F}, \mathrm{Dom}L)$ (Example 12G,(g)), and $L(v\otimes) = (Lv)\otimes = 0_{\mathbb{F} \to \mathrm{Cod}L}$. Since (iii) holds, $v\otimes = 0_{\mathbb{F} \to \mathrm{Dom}L}$, and hence $v = v \otimes 1 = 0$. Since $v \in \mathrm{Null}L$ was arbitrary, we have $\mathrm{Null}L = \{0\}$. by Corollary 14D, L is injective.

•(iv) *implies* (iii). Choose a linear space $\mathcal{U}$ such that $\mathcal{U}$ is not a zero-space and L is linearly left-cancellable with respect to $\mathcal{U}$. Choose $u \in \mathcal{U}^\times$. The $u\otimes \in \mathrm{Lin}(\mathbb{F}, \mathcal{U})$ is injective, by Proposition 11C; by •Corollary 21I,(a) we may choose a linear left-inverse $K \in \mathrm{Lin}(\mathcal{U}, \mathbb{F})$ of $u\otimes$. Now let $M \in \mathrm{Lin}(\mathbb{F}, \mathrm{Dom}L)$ be given, and assume that $LM = 0_{\mathbb{F} \to \mathrm{Cod}L}$. We have to show that $M = 0_{\mathbb{F} \to \mathrm{Dom}L}$. Now $LMK = 0_{\mathcal{U} \to \mathrm{Cod}L}$, and by the assumption on $\mathcal{U}$ we have $MK = 0_{\mathcal{U} \to \mathrm{Dom}L}$. Consequently $M = MK(u\otimes) = 0_{\mathbb{F} \to \mathrm{Dom}L}$, as was to be shown. ∎

•22F.R. PROPOSITION. *Let the linear mapping L be given. The following statements are equivalent.*

(i): *L is surjective.*

(ii): *L is a linear epimorphism.*

(iii): *L is linearly right-cancellable with respect to $\mathbb{F}$.*

(iv): *L is linearly right-cancellable with respect to some non-trivial linear space.*

Proof. The equivalence (i) $\Leftrightarrow$ (ii) is part of Proposition 22B.R. The implications (ii) $\Rightarrow$ (iii) $\Rightarrow$ (iv) are trivial. We shall prove •(iii) $\Rightarrow$ (ii) and (iv) $\Rightarrow$ (iii).

•(iii) *implies* (ii). Suppose that L is *not* a linear epimorphism. We may then choose a linear space $\mathcal{U}$ and $M \in \mathrm{Lin}(\mathrm{Cod}L, \mathcal{U})$ such that $ML = 0$ but $M \neq 0$. We choose $w \in \mathrm{Cod}L$ such that $Mw \neq 0$. The linear mapping $Mw\otimes \in \mathrm{Lin}(\mathbb{F}, \mathcal{U})$ is injective; by •Corollary 21I,(a) we may choose a linear left-inverse $H \in \mathrm{Lin}(\mathcal{U}, \mathbb{F})$ of $Mw\otimes$. Now $HM \in \mathrm{Lin}(\mathrm{Cod}L, \mathbb{F})$, and $HMw = HMw \otimes 1 = 1$, so that $HM \neq 0$. However, $HML = 0$. Therefore L is *not* linearly right-cancellable with respect to $\mathbb{F}$.

(iv) *implies* (iii). Choose a linear space $\mathcal{U}$ such that $\mathcal{U}$ is not a zero-space and L is linearly right-cancellable with respect to $\mathcal{U}$. Choose $u \in \mathcal{U}^\times$. Let $M \in \mathrm{Lin}(\mathrm{Cod}L, \mathbb{F})$

be given, and assume that $ML = 0$. We have to show that $M = 0$. Now $u \otimes ML = 0$, and by the assumption on $\mathcal{U}$ we have $u \otimes M = 0$. But $u\otimes$ is injective and therefore a linear monomorphism (Proposition 22B.L). Therefore $M = 0$. ∎

We now have available a linear analogue of *Basic Language*, •Theorem 36E.

•**22G. Theorem.** *Let the linear mappings $M : \mathcal{V} \to \mathcal{V}'$ and $N : \mathcal{W}' \to \mathcal{W}$ be given. For every $L \in \mathrm{Lin}(\mathcal{V}, \mathcal{W})$ there is at least one $L' \in \mathrm{Lin}(\mathcal{V}', \mathcal{W}')$ with $L = NL'M$ if and only if* $\mathrm{Null}L \supset \mathrm{Null}M$ *and* $\mathrm{Rng}L \subset \mathrm{Rng}N$. *There is at most one such L' for every $L \in \mathrm{Lin}(\mathcal{V}, \mathcal{W})$ if and only if either M is surjective and N is injective, or $\mathcal{V}'$ is a zero-space, or $\mathcal{W}'$ is a zero-space.*

Proof. The proof is entirely analogous to the proof of *Basic Language*, •Theorem 36E. It uses (22.5) and (22.6), •Corollary 21K, Theorem 22D, and – for the proof of the last part of the statement – •Propositions 22F.L and 22F.R. ∎

Chapter 3

LINEAR PRODUCTS AND COPRODUCTS

31. Linear products

In this section and the next we examine the formal analogues for linear spaces and linear mappings of the set-products and set-coproducts discussed in *Basic Language*, Section 46.

We begin with some additional facts about Cartesian products of families of linear spaces (Example 11D,(c); see also Examples 12G,(a),(c)). Let the family of linear spaces $(\mathcal{V}_i \mid i \in I)$ and the set D be given. In *Basic Language*, Proposition 44C we exhibited a bijection from $\underset{i\in I}{\bigtimes} \mathrm{Map}(D, \mathcal{V}_i)$ to $\mathrm{Map}(D, \underset{i\in I}{\bigtimes} \mathcal{V}_i)$ that served to identify these sets, as follows. For every family of mappings $(f_i \mid i \in I) \in \underset{i\in I}{\bigtimes} \mathrm{Map}(D, \mathcal{V}_i)$, the mapping in $\mathrm{Map}(D, \underset{i\in I}{\bigtimes} \mathcal{V}_i)$ denoted by the same symbol $(f_i \mid i \in I)$ is defined by the rule

$$(31.1) \qquad (f_i \mid i \in I)(x) := (f_i(x) \mid i \in I) \quad \text{for all } x \in D.$$

We now record that this bijection is linear, and that, if D is specialized to be a linear space, this bijection matches linear mappings with linear mappings.

31A. Proposition. *Let the family of linear spaces $(\mathcal{V}_i \mid i \in I)$ be given. For every set D and family of mappings $(f_i \mid i \in I) \in \underset{i\in I}{\bigtimes} \mathrm{Map}(D, \mathcal{V}_i)$, the only mapping $f \in \mathrm{Map}(D, \underset{i\in I}{\bigtimes} \mathcal{V}_i)$ satisfying $\pi_j f = f_j$ for all $j \in I$ is $f := (f_i \mid i \in I)$ defined by the rule (31.1). This formula describes a linear isomorphism from $\underset{i\in I}{\bigtimes} \mathrm{Map}(D, \mathcal{V}_i)$ to $\mathrm{Map}(D, \underset{i\in I}{\bigtimes} \mathcal{V}_i)$. If $\mathcal{U}$ is a linear space, then the corresponding linear isomorphism from $\underset{i\in I}{\bigtimes} \mathrm{Map}(\mathcal{U}, \mathcal{V}_i)$ to $\mathrm{Map}(\mathcal{U}, \underset{i\in I}{\bigtimes} \mathcal{V}_i)$ maps $\underset{i\in I}{\bigtimes} \mathrm{Lin}(\mathcal{U}, \mathcal{V}_i)$ onto $\mathrm{Lin}(\mathcal{U}, \underset{i\in I}{\bigtimes} \mathcal{V}_i)$;*

i.e., if $(L_i \mid i \in I) \in \bigtimes_{i \in I} \mathrm{Map}(\mathcal{U}, \mathcal{V}_i)$, *then* $(L_i \mid i \in I) \in \mathrm{Map}(\mathcal{U}, \bigtimes_{i \in I} \mathcal{V}_i)$ *is linear if and only if* L_i *is linear for every* $i \in I$.

   Proof. By reference to *Basic Language*, Proposition 44C and direct verification from the definitions of the linear-space structures involved. ∎

   Let the family of linear spaces $(\mathcal{V}_i \mid i \in I)$ be given. As an application of Proposition 31A we may define for each $j \in I$ the linear mapping $\bar{\sigma}_j \in \mathrm{Lin}(\mathcal{V}_j, \bigtimes_{i \in I} \mathcal{V}_i)$ by requiring

$$(31.2) \qquad \pi_i \bar{\sigma}_j := \begin{cases} 1_{\mathcal{V}_j} & \text{if } i = j \\[2mm] 0_{\mathcal{V}_j \to \mathcal{V}_i} & \text{if } i \in I \backslash \{j\}. \end{cases}$$

We note that in fact $\bar{\sigma}_j = (0, \cdot_j)$ (*Basic Language*, (44.1)).

   31B. COROLLARY. *Let the families of linear spaces* $(\mathcal{V}_i \mid i \in I)$ *and* $(\mathcal{W}_i \mid i \in I)$ *and the family of mappings* $(M_i \mid i \in I) \in \bigtimes_{i \in I} \mathrm{Map}(\mathcal{V}_i, \mathcal{W}_i)$ *be given. Then the mapping* $\bigtimes_{i \in I} M_i : \bigtimes_{i \in I} \mathcal{V}_i \to \bigtimes_{i \in I} \mathcal{W}_i$ *defined (according to Basic Language, Section 44) by requiring*

$$(31.3) \qquad \pi_j^{\mathcal{W}} \bigtimes_{i \in I} M_i := M_j \circ \pi_j^{\mathcal{V}} \quad \text{for all } j \in I,$$

is linear if and only if M_j *is linear for every* $i \in I$. *(*$(\pi_i^{\mathcal{V}} \mid i \in I)$ *and* $(\pi_i^{\mathcal{W}} \mid i \in I)$ *are the respective families of projections.)*

   Proof. If M_j is linear for every $j \in I$, then so is $M_j \circ \pi_j^{\mathcal{V}}$, and it follows from Proposition 31A and (31.3) that $\bigtimes_{i \in I} M_i$ is linear. Assume, conversely, that $\bigtimes_{i \in I} M_i$ is linear. Then (31.2) and (31.3) yield

$$M_j = M_j \circ (\pi_j^{\mathcal{V}} \bar{\sigma}_j) = \pi_j^{\mathcal{W}} (\bigtimes_{i \in I} M_i) \bar{\sigma}_j \quad \text{for all } j \in I.$$

Since all the mappings in the right-hand side are linear, so is M_j for every $j \in I$. ∎

   A **(linear) product of** a family of linear spaces $(\mathcal{V}_i \mid i \in I)$ is defined to be a linear space $\mathcal{P}$ together with a family of linear mappings $(p_i \mid i \in I) \in \bigtimes_{i \in I} \mathrm{Lin}(\mathcal{P}, \mathcal{V}_i)$ such that for every linear space $\mathcal{U}$ and every family of linear mappings $(L_i \mid i \in I) \in \bigtimes_{i \in I} \mathrm{Lin}(\mathcal{U}, \mathcal{V}_i)$ there is exactly one linear mapping $L : \mathcal{U} \to \mathcal{P}$ such that $L_i = p_i L$ for all $i \in I$. The linear space $\mathcal{P}$ is called the **product-space**, and for each $j \in I$ the linear mapping $p_j : \mathcal{P} \to \mathcal{V}_j$ is called the j**th projection**.

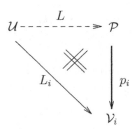

The first part of the next proposition asserts that a given family of linear spaces has "essentially" at most one product.

31C. PROPOSITION. (a): *Let linear products of the family of linear spaces $(\mathcal{V}_i \mid i \in I)$ be given, with respective product-spaces $\mathcal{P}$ and $\mathcal{P}'$ and respective families of projections $(p_i \mid i \in I)$ and $(p_i' \mid i \in I)$. Then the unique linear mappings $M : \mathcal{P} \to \mathcal{P}'$ and $M' : \mathcal{P}' \to \mathcal{P}$ that satisfy $p_i = p_i'M$ and $p_i' = p_i M'$ for all $i \in I$ are invertible, and each is the inverse of the other.*

(b): *Let a linear product of the family of linear spaces $(\mathcal{V}_i \mid i \in I)$ be given, with product-space $\mathcal{P}$ and family of projections $(p_i \mid i \in I)$. A given linear space $\mathcal{Q}$ and family of linear mappings $(q_i \mid i \in I) \in \bigtimes_{i \in I} \mathrm{Lin}(\mathcal{Q}, \mathcal{V}_i)$ are the product-space and family of projections of a linear product of $(\mathcal{V}_i \mid i \in I)$ if and only if the unique linear mapping $M : \mathcal{Q} \to \mathcal{P}$ that satisfies $q_i = p_i M$ for all $i \in I$ is invertible.*

Proof. This is entirely analogous to the proof of *Basic Language*, Proposition 46A. ∎

We now show that every family of linear spaces has a linear product.

31D. PROPOSITION. *Let the family of linear spaces $(\mathcal{V}_i \mid i \in I)$ be given. Then the Cartesian product $\bigtimes_{i \in I} \mathcal{V}_i$ with the family of projections $(\pi_i \mid i \in I)$ is a linear product of $(\mathcal{V}_i \mid i \in I)$.*

Proof. Examples 11D,(c) and 2G,(a), and Proposition 31A. ∎

The special linear product described by Proposition 31D may be called the **standard (linear) product of** the family of linear spaces $(\mathcal{V}_i \mid i \in I)$.

31E. EXAMPLE. Let the set D and the linear space $\mathcal{V}$ be given. Then the linear space $\mathrm{Map}(D, \mathcal{V})$ and the family of linear mappings $\mathrm{ev}^{\mathrm{Map}(D,\mathcal{V})}$ (Examples 11D,(d) and 12G,(b)) are the product-space and the family of projections of a linear product of the family $(\mathcal{V} \mid x \in D)$. ∎

We next demonstrate the intimate relationship between linear products and set-products.

31F. PROPOSITION. *Let the family of linear spaces $(\mathcal{V}_i \mid i \in I)$ be given.*

(a): *If the linear space $\mathcal{P}$ with the family of linear mappings $(p_i \mid i \in I)$ is a linear product of $(\mathcal{V}_i \mid i \in I)$, then the set $\mathcal{P}$ with the family of mappings $(p_i \mid i \in I)$ is a set-product of the family of sets $(\mathcal{V}_i \mid i \in I)$.*

(b): *If the set $\mathcal{P}$ with the family of mappings $(p_i \mid i \in I)$ is a set-product of the family of sets $(\mathcal{V}_i \mid i \in I)$, there is exactly one linear-space structure on $\mathcal{P}$ such that*

p_i is linear for every $i \in I$. Endowed with this linear-space structure, $\mathcal{P}$ with the family of linear mappings $(p_i \mid i \in I)$ is a linear product of the family of linear spaces $(\mathcal{V}_i \mid i \in I)$.

Proof. Proof of (a). By Proposition 31D and 31C,(a), there is a unique linear isomorphism $M : \mathcal{P} \to \underset{i \in I}{\times} \mathcal{V}_i$ such that

$$(31.4) \qquad\qquad p_j = \pi_j \circ M \quad \text{for all } j \in I.$$

Now the set $\underset{i \in I}{\times} \mathcal{V}_i$ with $(\pi_i \mid i \in I)$ is a set-product of the family of sets $(\mathcal{V}_i \mid i \in I)$ (*Basic Language*, Proposition 46B), and $M : \mathcal{P} \to \underset{i \in I}{\times} \mathcal{V}_i$ is bijective and satisfies (31.4). By *Basic Language*, Proposition 46A,(b), the set $\mathcal{P}$ with $(p_i \mid i \in I)$ is a set-product of the family of sets $(\mathcal{V}_i \mid i \in I)$.

Proof of (b). We set $\mathcal{W} := \underset{i \in I}{\times} \mathcal{V}_i$. By *Basic Language*, Propositions 46B and 46A,(a), there is a unique bijection $g : \mathcal{P} \to \mathcal{W}$ such that

$$(31.5) \qquad\qquad p_j = \pi_j \circ g \quad \text{for all } j \in I.$$

Suppose a linear-space structure is given on $\mathcal{P}$ by the prescription of $0^{\mathcal{P}}$, $\mathrm{add}^{\mathcal{P}}$, $\mathrm{opp}^{\mathcal{P}}$, $\mathrm{mult}^{\mathcal{P}}$. Regarding $\mathcal{W} = \underset{i \in I}{\times} \mathcal{V}_i$ as endowed with its linear-space structure (Example 11D,(c)), we claim that p_j is linear for all $j \in I$ if and only if g is linear. Indeed, for every $j \in I$,

$$p_j \circ \mathrm{add}^{\mathcal{P}} = \pi_j \circ g \circ \mathrm{add}^{\mathcal{P}}$$

$$\mathrm{add}^{\mathcal{V}_j} \circ (p_j \times p_j) = \mathrm{add}^{\mathcal{V}_j} \circ ((\pi_j \circ g) \times (\pi_j \circ g)) = \mathrm{add}^{\mathcal{V}_j} \circ (\pi_j \times \pi_j) \circ (g \times g) = \pi_j \circ \mathrm{add}^{\mathcal{W}} \circ (g \times g),$$

and hence p_j is additive for all $j \in I$ if and only if

$$\pi_j \circ g \circ \mathrm{add}^{\mathcal{P}} = \pi_j \circ \mathrm{add}^{\mathcal{W}} \circ (g \times g) \quad \text{for all } j \in I;$$

and this holds if and only if $g \circ \mathrm{add}^{\mathcal{P}} = \mathrm{add}^{\mathcal{W}} \circ (g \times g)$, i.e., g is additive. The proof for the homogeneity is similar, but simpler.

To complete the proof of the first assertion, it now suffices to observe that, since g is bijective, there is obviously exactly one linear-space structure on $\mathcal{P}$ that makes g linear; it is given by

$$0^{\mathcal{P}} := g^{\leftarrow}(0^{\mathcal{W}}) \qquad \mathrm{add}^{\mathcal{P}} := g \circ \mathrm{add}^{\mathcal{W}} \circ (g^{\leftarrow} \times g^{\leftarrow}) \qquad \mathrm{opp}^{\mathcal{P}} := g \circ \mathrm{opp}^{\mathcal{W}} \circ g^{\leftarrow}$$

$$\mathrm{mult}^{\mathcal{P}}_s := g \circ \mathrm{mult}^{\mathcal{W}}_s \circ g^{\leftarrow} \qquad \text{for all } s \in \mathbb{F}.$$

When $\mathcal{P}$ is endowed with this linear-space structure, then g is linear, and p_j is linear for every $j \in J$; it follows from Propositions 31D and 31C, (b) and (31.5) that the linear space $\mathcal{P}$ with $(p_i \mid i \in I)$ is a linear product of the family of linear spaces $(\mathcal{V}_i \mid i \in I)$. ∎

31G. Proposition. *Let a linear product of the family of linear spaces $(\mathcal{V}_i \mid i \in I)$ be given, with product-space $\mathcal{P}$ and family of projections $(p_i \mid i \in I)$.*

(a): $\forall u, v \in \mathcal{P}, \quad u = v \quad \Leftrightarrow \quad (\forall i \in I, \ p_i u = p_i v)$.

(b): p_j *is linearly right-invertible, hence surjective, for every $j \in I$.*

Proof. (a) follows from Proposition 31F,(a) and *Basic Language*, Proposition 46D,(a). To prove (b), we define, for every $j \in I$, the linear mapping $\bar{s}_j \in \mathrm{Lin}(\mathcal{V}_j, \mathcal{P})$ by requiring

$$(31.6) \qquad p_i \bar{s}_j := \begin{cases} 1_{\mathcal{V}_j} & \text{if } i = j \\[2ex] 0_{\mathcal{V}_j \to \mathcal{V}_i} & \text{if } i \in I \backslash \{j\}. \end{cases}$$

Then $\bar{s}_j$ is a linear right-inverse of p_j for each $j \in I$. ∎

32. Linear coproducts

Before dealing with (abstract) linear coproducts, defined in analogy with set-coproducts, we obtain additional information on direct sums of families of linear spaces, as defined in Example 13G,(e).

Let the family of linear spaces $(\mathcal{V}_i \mid i \in I)$ be given, and consider, for each $j \in I$, the linear mapping $\bar{\sigma}_j \in \mathrm{Lin}(\mathcal{V}_j, \bigtimes_{i \in I} \mathcal{V}_i)$ defined by (31.2). Obviously, $\mathrm{Supp}(\bar{\sigma}_j u) \subset \{j\}$ for all $u \in \mathcal{V}_j$, and so $\mathrm{Rng}\bar{\sigma}_j \subset \bigtimes_{i \in I} \mathcal{V}_i$ for every $j \in I$. We define $\sigma_j \in \mathrm{Lin}(\mathcal{V}_j, \bigoplus_{i \in I} \mathcal{V}_i)$ to be the linear mapping obtained from $\bar{\sigma}_j$ by adjustment of the codomain to $\bigoplus_{i \in I} \mathcal{V}_i$, for every $j \in I$.

32A. LEMMA. *Let the family of linear spaces $(\mathcal{V}_i \mid i \in I)$ be given. Then*

$$(32.1) \qquad \pi_i(\sigma_j u) := \begin{cases} u & \text{if } i = j \\[2mm] 0 & \text{if } i \in I \setminus \{j\} \end{cases} \qquad \text{for all } j \in I \text{ and } u \in \mathcal{V}_j;$$

and

$$(32.2) \qquad \mathrm{Supp}(\sigma_i \pi_i v \mid i \in I) \text{ is finite and } \sum_{i \in I} \sigma_i \pi_i v = v \quad \text{for all } v \in \bigoplus_{i \in I} \mathcal{V}_i.$$

Proof. (32.1) is an immediate consequence of (31.2). Let $v \in \bigoplus_{i \in I} \mathcal{V}_i$ be given. Then $\mathrm{Supp}(\sigma_i \pi_i v \mid i \in I) \subset \mathrm{Supp}(\pi_i v \mid i \in I) = \mathrm{Supp}\, v$, which is finite. By Proposition 12B and (32.1) we have

$$\pi_j \Big(\sum_{i \in I} \sigma_i \pi_i v \Big) = \sum_{i \in I} \pi_j(\sigma_j \pi_i v) = \pi_j v \quad \text{for all } j \in I,$$

and therefore $\sum_{i \in I} \sigma_i \pi_i v = v$. ∎

32B. PROPOSITION. *Let the family of linear spaces $(\mathcal{V}_i \mid i \in I)$ and the linear space $\mathcal{U}$ be given. For every family of linear mappings $(L_i \mid i \in I) \in \bigtimes_{i \in I} \mathrm{Lin}(\mathcal{V}_i, \mathcal{U})$ there is exactly one linear mapping $L \in \bigoplus_{i \in I} \mathcal{V}_i \to \mathcal{U}$ such that $L_j = L\sigma_j$ for all $j \in I$; it is defined by*

$$(32.3) \qquad Lv := \sum_{i \in I} L_i \pi_i v = \sum_{i \in I} L_i v_i \quad \text{for all } v \in \bigoplus_{i \in I} \mathcal{V}_i.$$

Proof. Let the family of linear mappings $(L_i \mid i \in I) \in \bigtimes_{i \in I} \mathrm{Lin}(\mathcal{V}_i, \mathcal{U})$ be given.
If $L \in \mathrm{Lin}\bigoplus_{i \in I} \mathcal{V}_i, \mathcal{U})$ satisfies $L_j = L\sigma_j$ for all $j \in I$, then (32.2) and Proposition 12B
yield

$$Lv = L\sum_{i \in I} \sigma_i \pi_i v = \sum_{i \in I} L\sigma_i \pi_i v = \sum_{i \in I} L_i \pi_i v \quad \text{for all } v \in \bigoplus_{i \in I} \mathcal{V}_i.$$

Conversely, define the mapping $L : \bigoplus_{i \in I} \mathcal{V}_i \to \mathcal{U}$ by setting $L(v) := \sum_{i \in I} L_i \pi_i v$ for
all $v \in \bigoplus_{i \in I} \mathcal{V}_i$. Then L is linear, as can be verified directly, and for every $j \in I$ we
have, by (32.1),

$$L\sigma_j u = \sum_{i \in I} L_i \pi_i (\sigma_j u) = L_j u \quad \text{for all } u \in \mathcal{V}_j,$$

so that $L\sigma_j = L_j$. ∎

A **(linear) coproduct of** a family of linear spaces $(\mathcal{V}_i \mid i \in I)$ is defined to be a
linear space $\mathcal{S}$ together with a family of linear mappings $(s_i \mid i \in I) \in \bigtimes_{i \in I} \mathrm{Lin}(\mathcal{V}_i, \mathcal{S})$
such that for every linear space $\mathcal{U}$ and every family of linear mappings
$(L_i \mid i \in I) \in \bigtimes_{i \in I} \mathrm{Lin}(\mathcal{V}_i, \mathcal{U})$ there is exactly one linear mapping $L : \mathcal{S} \to \mathcal{U}$ such that
$L_i = Ls_i$ for all $i \in I$. The linear space $\mathcal{S}$ is called the **coproduct-space**, and for
each $j \in I$, the linear mapping $s_j : \mathcal{V}_j \to \mathcal{S}$ is called the **jth insertion**.

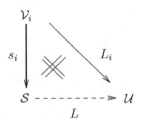

32C. PROPOSITION. (a): *Let linear coproducts of the family of linear spaces*
$(\mathcal{V}_i \mid i \in I)$ *be given, with respective coproduct-spaces* $\mathcal{S}$ *and* $\mathcal{S}'$ *and respective families of insertions* $(s_i \mid i \in I)$ *and* $(s'_i \mid i \in I)$. *Then the unique linear mappings*
$M : \mathcal{S}' \to \mathcal{S}$ *and* $M' : \mathcal{S} \to \mathcal{S}'$ *that satisfy* $s_i = Ms'_i$ *and* $s'_i = M's_i$ *for all* $i \in I$ *are
invertible, and each is the inverse of the other.*

(b): *Let a linear coproduct of the family of linear spaces* $(\mathcal{V}_i \mid i \in I)$ *be given,
with coproduct-spaces* $\mathcal{S}$ *and family of insertions* $(s_i \mid i \in I)$. *A given linear space* $\mathcal{T}$
and family of linear mappings $(t_i \mid i \in I) \in \bigtimes_{i \in I} \mathrm{Lin}(\mathcal{V}_i, \mathcal{T})$ *are the coproduct-space
and family of insertions of a linear coproduct of* $(\mathcal{V}_i \mid i \in I)$ *if and only if the unique
linear mapping* $M : \mathcal{S} \to \mathcal{T}$ *that satisfies* $t_i = Ms_i$ *for all* $i \in I$ *is invertible.*

Proof. This is analogous to the proof of Proposition 31C, and hence to the proof of *Basic Language*, Proposition 46A. ∎

32D. PROPOSITION. *Let the family of linear spaces $(V_i \mid i \in I)$ be given. Then the direct sum $\bigoplus_{i \in I} V_i$ with the family $(\sigma_i \mid \in I)$ is a linear coproduct of $(V_i \mid i \in I)$.*

Proof. Proposition 32B. ∎

The special linear coproduct described by Proposition 32D may be called the **standard (linear) coproduct of** the family of linear spaces $(V_i \mid i \in I)$. In the next section we shall discuss other important linear coproducts.

32E. PROPOSITION. *Let a linear coproduct of the family of linear spaces $(V_i \mid i \in I)$ be given, with coproduct-space S and family of insertions $(s_i \mid i \in I)$. For each $j \in I$ define $\bar{p}_j \in \mathrm{Lin}(S, V_j)$ by requiring*

$$(32.4) \qquad \bar{p}_j s_i := \begin{cases} 1_{V_j} & \text{if } i = j \\ 0_{V_i \to V_j} & \text{if } i \in I \backslash \{j\}. \end{cases}$$

For each $j \in I$, s_j is linearly left-invertible, hence injective, and $\bar{p}_j$ is linearly right-invertible, hence surjective. Moreover,

$$(32.5) \qquad \mathrm{Supp}(s_i \bar{p}_i v \mid i \in I) \text{ is finite, and } \sum_{i \in I} s_i \bar{p}_i v = v \text{ for all } v \in S.$$

If, in particular, $\mathrm{Supp}(V_i \mid i \in I)$ is finite, then $\mathrm{Supp}(s_i \bar{p}_i \mid i \in I)$ is finite and

$$(32.6) \qquad \sum_{i \in I} s_i \bar{p}_i = 1_S.$$

Proof. We could derive this from Lemma 32A by way of Propositions 32D and 32C, but it is instructive to give a direct proof. We set

$$W := \{v \in S \mid \mathrm{Supp}(s_i \bar{p}_i v \mid i \in I) \text{ is finite}\}$$
$$U := \{v \in W \mid \sum_{i \in I} s_i \bar{p}_i v = v\}.$$

Now W is a subspace of S, since it is the pre-image under the linear mapping $(s_i \bar{p}_i \mid i \in I) : S \to S^I$ of the subspace $S^{(I)}$ of S^I. But then U is a subspace of W, since it is the null-space of $AB - 1_{W \subset S}$, where $B := (s_i \bar{p}_i \mid i \in I)|^{S^{(I)}}$ and $A := (a \mapsto \sum_I a) : S^{(I)} \to S$ are linear. We are to show that $U = S$.

For every $j \in I$ and $u \in V_j$ we have $s_i \bar{p}_i s_j u = 0$ for all $i \in I \backslash \{j\}$, and $\sum_{i \in I} s_i \bar{p}_i s_j u = s_j \bar{p}_j s_j u = s_j u$, by (32.4), and therefore $s_j u \in U$. Thus $\mathrm{Rng} s_j \subset U$ for all $j \in I$.

By the definition of linear coproduct, there is exactly one linear mapping $L : \mathcal{S} \to \mathcal{U}$ such that $s_j|^{\mathcal{U}} = Ls_j$ for all $j \in I$. But then

$$1_{\mathcal{S}} s_j = s_j = (s_j|^{\mathcal{U}})|^{\mathcal{S}} = (L|^{\mathcal{S}}) s_j \quad \text{for all } j \in I.$$

Again by the definition of linear coproduct, this implies $1_{\mathcal{S}} = L|^{\mathcal{S}}$; and therefore $\mathcal{S} = \text{Rng} 1_{\mathcal{S}} = \text{Rng}(L|^{\mathcal{S}}) = \text{Rng} L \subset \mathcal{U} \subset \mathcal{S}$. We conclude that $\mathcal{U} = \mathcal{S}$, as was to be shown. ∎

32F. Corollary. *Let a linear coproduct of the family of linear spaces* $(\mathcal{V}_i \mid i \in I)$ *be given, with coproduct-spaces* $\mathcal{S}$ *and family of insertions* $(s_i \mid i \in I)$. *Then* $\mathcal{S} = \sum_{i \in I} \text{Rng} s_i$.

32G. Remark. In contrast with the results described in Proposition 31F for linear products, the coproduct-space and family of insertions of a *linear coproduct* of a family of *linear spaces* $(\mathcal{V}_i \mid i \in I)$ are *not* the coproduct-set and family of insertions of a *set-coproduct* of the family of sets $(\mathcal{V}_i \mid i \in I)$, unless I is a singleton: in a linear coproduct the coproduct-space is never empty, and 0 is in the range of every insertion; in a set-coproduct, the coproduct-set is empty if the index set is empty, and the family of ranges of the insertions is disjoint. ∎

32H. Proposition. *Let both a linear product and a linear coproduct of a family of linear spaces* $(\mathcal{V}_i \mid i \in I)$ *be given, with product-space* $\mathcal{P}$ *and family of projections* $(p_i \mid i \in I)$, *and coproduct-space* $\mathcal{S}$ *and family to insertions* $(s_i \mid i \in I)$, *respectively. Let* $(\bar{p}_i \mid i \in I) \in \underset{i \in I}{\bigtimes} \text{Lin}(\mathcal{S}, \mathcal{V}_i)$ *and* $(\bar{s}_i \mid i \in I) \in \underset{i \in I}{\bigtimes} \text{Lin}(\mathcal{V}_i, \mathcal{P})$ *be defined by (32.4) and (31.6), respectively.*

(a): *There is exactly one linear mapping* $K : \mathcal{S} \to \mathcal{P}$ *that satisfies*

$$p_j K s_i = \begin{cases} 1_{\mathcal{V}_j} & if \ i = j \\ \\ 0_{\mathcal{V}_i \to \mathcal{V}_j} & if \ i \neq j \end{cases} \quad for \ all \ i, j \in I;$$

it is the only linear mapping that satisfies $p_j K = \bar{p}_j$ *for all* $j \in I$ *and also the only linear mapping that satisfies* $K s_j = \bar{s}_j$ *for all* $j \in I$. *Moreover,* K *is injective.*

(b): *Assume that* $\text{Supp}(\mathcal{V}_i \mid i \in I)$ *is finite (this is the case, in particular, when* I *is finite). Then* K *is invertible;* $\mathcal{S}$ *is the product-space and* $(\bar{p}_i \mid i \in I)$ *the family of projections of a linear product of* $(\mathcal{V}_i \mid i \in I)$; *and* $\mathcal{P}$ *is the coproduct-space and* $(\bar{s}_i \mid i \in I)$ *the family of insertions of a linear coproduct of* $(\mathcal{V}_i \mid i \in I)$.

Proof. Proof of (a). Except for the injectivity of K, the assertion (a) is an immediate consequence of the definitions of linear product and linear coproduct and (32.4) and (31.6). By (32.5) we have

$$v = \sum_{i \in I} s_i \bar{p}_i v = \sum_{i \in I} s_i p_i K v = 0 \quad \text{for all } v \in \text{Null} K.$$

Hence $\text{Null} K = \{0\}$. By Corollary 14D, K is injective.

Proof of (b). We assume that $\mathrm{Supp}(\mathcal{V}_i \mid i \in I)$ is finite. Let the linear space $\mathcal{U}$ and the family of linear mappings $(L_i \mid i \in I) \in \underset{i \in I}{\LARGE\times} \mathrm{Lin}(\mathcal{U}, \mathcal{V}_i)$ be given. If $\bar{L} \in \mathrm{Lin}(\mathcal{U}, \mathcal{S})$ satisfies $\bar{p}_j \bar{L} = L_j$ for all $j \in I$, then by (32.6) and Proposition 12B and Remark 13I we have

$$\bar{L} = \Big(\sum_{i \in I} s_i \bar{p}_i \Big) \bar{L} = \sum_{i \in I} s_i \bar{p}_i \bar{L} = \sum_{i \in I} s_i L_i.$$

Conversely, we have $\mathrm{Supp}(s_i L_i \mid i \in I) \subset \mathrm{Supp}(\mathcal{V}_i \mid i \in I)$, so that the former set is finite, and we may *define* $\bar{L} := \sum_{i \in I} s_i L_i \in \mathrm{Lin}(\mathcal{U}, \mathcal{S})$. Then by (32.4) and Proposition 12B and Remark 13I,

$$\bar{p}_j \bar{L} = \bar{p}_j \sum_{i \in I} s_i L_i = \sum_{i \in I} \bar{p}_j s_i L_i = L_j \quad \text{for all } j \in I.$$

Thus $\sum_{i \in I} s_i L_i$ is the only $\bar{L} \in \mathrm{Lin}(\mathcal{U}, \mathcal{S})$ such that $\bar{p}_j \bar{L} = L_j$ for all $j \in I$. This proves that $\mathcal{S}$ is the product-space and $(\bar{p}_i \mid i \in I)$ the family of projections of a linear product of $(\mathcal{V}_i \mid i \in I)$.

By Proposition 31C,(b), applied to $\mathcal{S}$, $(\bar{p}_i \mid i \in I)$, K instead of $\mathcal{Q}$, $(q_i \mid i \in I)$, M, it follows that K is invertible. Now $K^{-1} \bar{s}_j = K^{-1} K s_j = s_j$ for all $j \in I$. We may therefore apply Proposition 32C,(b) to $\mathcal{P}$, $(\bar{s}_i \mid i \in I)$, K^{-1} instead of $\mathcal{T}$, $(t_i \mid i \in I)$, M, and conclude that $\mathcal{P}$ is the coproduct-space and $\bar{s}_i \mid i \in I)$ the family of insertions of a linear coproduct of $(\mathcal{V}_i \mid i \in I)$. ∎

32I. REMARKS. (a): An alternative proof of Proposition 32H can be obtained from consideration of the standard linear product and coproduct, and the observation that

$$\bigoplus_{i \in I} \mathcal{V}_i = \underset{i \in I}{\LARGE\times} \mathcal{V}_i \text{ (and hence } \sigma_j = \bar{\sigma}_j \text{ and } \bar{\pi}_j = \pi_j \text{ for all } j \in I) \text{ if } \mathrm{Supp}(\mathcal{V}_i \mid i \in I) \text{ is}$$

finite (Example 13G,(e)).

•(b): From (a) it can be seen that the condition "$\mathrm{Supp}(\mathcal{V} \mid i \in I)$ is finite" in Proposition 32H,(b) is necessary for each of the conclusions, since it is •necessary for

$$\bigoplus_{i \in I} \mathcal{V}_i = \underset{i \in I}{\LARGE\times} \mathcal{V}_i \text{ (Example 13G,(e))}.$$

(c): If $\mathcal{V}_i = \mathcal{V}$ for all $i \in I$, the condition "$\mathrm{Supp}(\mathcal{V}_i \mid i \in I)$ is finite" may be replaced in Proposition 32H,(b) by the necessary and sufficient condition "I is finite or $\mathcal{V}$ is a zero-space". ∎

33. Decompositions

Let the linear space $\mathcal{V}$ be given. A family $(\mathcal{U}_i \mid i \in I)$ of subspaces of $\mathcal{V}$ is called a **decomposition of** $\mathcal{V}$ if for every linear space $\mathcal{W}$ and every family of linear mappings $(L_i \mid i \in I) \in \underset{i \in I}{\text{\Large\times}} \text{Lin}(\mathcal{U}_i, \mathcal{W})$ there is exactly one linear mapping $L : \mathcal{V} \to \mathcal{W}$ such that $L|_{\mathcal{U}_i} = L_i$ for all $i \in I$; in other words, if $\mathcal{V}$ is the coproduct space and the family $(1_{\mathcal{U}_i \subset \mathcal{V}} \mid i \in I)$ of inclusion mappings is the family of insertions of a linear coproduct of $(\mathcal{U}_i \mid i \in I)$.

This definition stresses the most important use of decompositions; they can, however, be characterized in a more geometric fashion, which exhibits them as a generalization of supplementary pairs of subspaces.

33A. Theorem. *Let the linear space $\mathcal{V}$ and the family $(\mathcal{U}_i \mid i \in I)$ of subspaces of $\mathcal{V}$ be given. The following statements are equivalent.*

(i): *$(\mathcal{U}_i \mid i \in I)$ is a decomposition of $\mathcal{V}$.*

(ii): *The linear mapping $S : \bigoplus_{i \in I} \mathcal{U}_i \to \mathcal{V}$ defined by the rule*

$$(33.1) \qquad Su := \sum_I u \quad \text{for all } u \in \bigoplus_{i \in I} \mathcal{U}_i$$

is invertible.

(iii): *$(\mathcal{U}_i \mid i \in I)$ satisfies*

$$(33.2) \qquad \sum_{i \in I} \mathcal{U}_i = \mathcal{V}$$

$$(33.3) \qquad \mathcal{U}_j \cap \sum_{i \in I \setminus \{j\}} \mathcal{U}_i = \{0\} \quad \text{for all } j \in I.$$

(iv): *If $I = \emptyset$, then $\mathcal{V} = \{0\}$; if $I \neq \emptyset$, then $\mathcal{U}_j$ is a supplement of $\sum_{i \in I \setminus \{j\}} \mathcal{U}_i$ in $\mathcal{V}$ for all $j \in I$.*

Proof. (i) *is equivalent to* (ii). Using (33.1) and (32.1) we have

$$S\sigma_j v = \sum_I \sigma_j v = \sum_{i \in I} \pi_i(\sigma_j v) = v \quad \text{for all } j \in I \text{ and } v \in \mathcal{U}_j,$$

and therefore $S\sigma_j = 1_{\mathcal{U}_j \subset \mathcal{V}}$ for all $j \in I$. The asserted equivalence now follows from Proposition 32C,(a),(b), applied to the standard coproduct of $(\mathcal{U}_i \mid i \in I)$ on the one hand, and to the space $\mathcal{V}$ and the family $(1_{\mathcal{U}_i \subset \mathcal{V}} \mid i \in I)$ on the other.

(ii) *is equivalent to* (iii). We observe that $\bigoplus_{i \in I} \mathcal{U}_i = \mathcal{V}^{(I)} \cap \underset{i \in I}{\text{\Large\times}} \mathcal{U}_i$, and therefore the linear mapping S defined by (33.1) satisfies $\text{Rng} S = \sum_{i \in I} \mathcal{U}_i$. Therefore S is surjective if and only if (33.2) holds.

By Corollary 14D, S fails to be injective if and only if Null$S \neq \{0\}$. This is the case if and only if we may choose $j \in I$ and $w \in \mathcal{U}_j^\times$ such that $w + \sum_{i \in I \setminus \{j\}} u_i = 0$ for a suitable $u \in \bigoplus_{i \in I \setminus \{j\}} \mathcal{U}_i$; equivalently, such that $w \in \sum_{i \in I \setminus \{j\}} \mathcal{U}_i$. We conclude that S fails to be injective if and only if $\mathcal{U}_j \cap \sum_{i \in I \setminus \{j\}} \mathcal{U}_i \neq \{0\}$ for some $j \in I$; hence S is injective if and only if (33.3) holds.

(iii) *is equivalent to* (iv). This is obvious once we observe that $\sum_{i \in I} \mathcal{U}_i = \mathcal{U}_j + \sum_{i \in I \setminus \{j\}} \mathcal{U}_i$ for each $j \in I$. ∎

33B. Corollary. *Let the linear space $\mathcal{V}$ and the family $(\mathcal{U}_i \mid i \in I)$ of subspaces of $\mathcal{V}$ be given. The following statements are equivalent:*

(i): *$(\mathcal{U}_i \mid i \in I)$ is a decomposition of $\sum_{i \in I} \mathcal{U}_i$.*

(ii): *The linear mapping $S : \bigoplus_{i \in I} \mathcal{U}_i \to \mathcal{V}$ defined by (33.1) is injective.*

(iii): *$(\mathcal{U}_i \mid i \in I)$ satisfies (33.3).*

33C. Corollary. *Let the linear space $\mathcal{V}$ and the subspaces $\mathcal{U}$ and $\mathcal{U}'$ of $\mathcal{V}$ be given. Then the list $(\mathcal{U},\mathcal{U}')$ is a decomposition of $\mathcal{V}$ if and only if $\mathcal{U}$ and $\mathcal{U}'$ are supplementary in $\mathcal{V}$.*

33D. Remarks. (a): In the mathematical literature, a decomposition is often referred to as a *direct decomposition*. The fact that $(\mathcal{U}_i \mid i \in I)$ is a decomposition of $\mathcal{V}$ is often expressed by saying that $\mathcal{V}$ is the *internal direct sum of* $(\mathcal{U}_i \mid i \in I)$. It is not unusual to find this phrase without the adjective "internal", and even the notation $\mathcal{V} = \bigoplus_{i \in I} \mathcal{U}_i$ is used. This amounts to using the linear isomorphism S defined by (33.1) to identify $\mathcal{V}$ with the direct sum of $(\mathcal{U}_i \mid i \in I)$ as defined in Example 13G,(e); we do not regard this identification as advisable.

(b): Condition (33.3) is a generalization of disjunctness of a pair of subspaces. A family $(\mathcal{U}_i \mid i \in I)$ that satisfies (33.3), and hence the equivalent conditions recited in Corollary 33B, is therefore said to be **disjunct**. *Warning*: In general it is not sufficient for $(\mathcal{U}_i \mid i \in I)$ to be disjunct that $\mathcal{U}_i \cap \mathcal{U}_j = \{0\}$ for all $i, j \in I$ with $i \neq j$. ∎

Let the linear space $\mathcal{V}$ and the decomposition $(\mathcal{U}_i \mid i \in I)$ be given. For each $j \in I$ we define $P_j \in \mathrm{Lin}(\mathcal{V},\mathcal{U}_j)$ by the rule

(33.4)
$$P_j|_{\mathcal{U}_i} := \begin{cases} 1_{\mathcal{U}_j} & \text{if } i = j \\ 0_{\mathcal{U}_i \to \mathcal{U}_j} & \text{if } i \in I \setminus \{j\}. \end{cases} \qquad \text{for all } j \in I.$$

The family $(P_i \mid i \in I)$ is just the special case, for the coproduct associated with the decomposition, of the family $(\bar{p}_i \mid i \in I)$ defined by (32.4). When $\mathrm{Supp}(\mathcal{U}_i \mid i \in I)$ is finite, Proposition 32H shows that this is the family of projections of a *product* of $(\mathcal{U}_i \mid i \in I)$ whose product space is $\mathcal{V}$. By poetic license, P_j is called the *j*th **projection of** the decomposition $(\mathcal{U}_i \mid i \in I)$ for every $j \in I$, even when $\text{Supp}(\mathcal{U}_i \mid i \in I)$ is not finite.

We also define the family $(E_i \mid i \in I) \in (\text{Lin}\mathcal{V})^I$ by the rule

$$(33.5) \qquad E_j := P_j|^\mathcal{V} = 1_{\mathcal{U}_j \subset \mathcal{V}} P_j \quad \text{for all } j \in I.$$

It follows easily from (33.5) and (33.4) that

$$(33.6) \qquad E_j E_i = \begin{cases} E_j & \text{if } i = j \\ 0 & \text{if } i \neq j. \end{cases} \quad \text{for all } i, j \in I.$$

Since $E_j E_j = E_j$, E_j is called the *j*th **idempotent of** the decomposition $(\mathcal{U}_i \mid i \in I)$ for every $j \in I$.

33E. PROPOSITION. *Let the linear space $\mathcal{V}$ and the decomposition $(\mathcal{U}_i \mid i \in I)$ of $\mathcal{V}$ be given, with families of projections and of idempotents $(P_i \mid i \in I)$ and $(E_i \mid i \in I)$, respectively. Then* $\text{Rng}E_j = \text{Rng}P_j = \mathcal{U}_j$ *and* $\text{Null}E_j = \text{Null}P_j = \sum_{i \in I \setminus \{j\}} \mathcal{U}_i$ *for all $j \in I$.*

Proof. Let $j \in I$ be given. By (33.4) we have $P_j 1_{\mathcal{U}_j \subset \mathcal{V}} = 1_{\mathcal{U}_j}$. By Corollary 21E,(c), P_j is surjective (so that $\text{Rng}E_j = \text{Rng}P_j = \mathcal{U}_j$), and $\text{Null}E_j = \text{Null}P_j$ is a supplement of $\mathcal{U}_j = \text{Rng}1_{\mathcal{U}_j \subset \mathcal{V}}$ in $\mathcal{V}$. By Theorem 33A, $\sum_{i \in I \setminus \{j\}} \mathcal{U}_i$ is also a supplement of $\mathcal{U}_j$ in $\mathcal{V}$. But $\mathcal{U}_i \subset \text{Null}P_j$ for all $i \in I \setminus \{j\}$, by (33.4). It follows from this by Proposition 13R that $\sum_{i \in I \setminus \{j\}} \mathcal{U}_i \subset \text{Null}P_j$. By Proposition 15D, these supplements of $\mathcal{U}_j$ in $\mathcal{V}$ must be one and the same. ∎

33F. PROPOSITION. *Let the linear space $\mathcal{V}$, the family $(\mathcal{U}_i \mid i \in I)$ of subspaces of $\mathcal{V}$, and the family of linear mappings $(E_i \mid i \in I)$ in $\text{Lin}\mathcal{V}$ be given. Then $(\mathcal{U}_i \mid i \in I)$ is a decomposition of $\mathcal{V}$ and $(E_i \mid i \in I)$ is its family of idempotents if and only if*

$$(33.7) \qquad \text{Rng}E_i = \mathcal{U}_i \quad \text{for all } i \in I,$$

$$(33.8) \qquad E_j E_i = \delta_{i,j}^I E_i \quad \text{for all } i, j \in I,$$

$$(33.9) \qquad \text{Supp}(E_i v \mid i \in I) \text{ is finite and } \sum_{i \in I} E_i v = v \text{ for all } v \in \mathcal{V}.$$

If $\text{Supp}(\mathcal{U}_i \mid i \in I)$ *is finite, (33.9) may be replaced by*

$$(33.10) \qquad\qquad \sum_{i\in I} E_i = 1_{\mathcal{V}}.$$

Proof. Proof of the "only if" part. We assume that $(\mathcal{U}_i \mid i \in I)$ is a decomposition of $\mathcal{V}$ and that $(E_i \mid i \in I)$ is its family of idempotents. Then (33.7) follows from Proposition 33E, and (33.8) from (33.6). On account of (33.5), we obtain (33.9) from Proposition 32E, applied to $(\mathcal{U}_i \mid i \in I)$, $\mathcal{V}, (1_{\mathcal{U}_i \subset \mathcal{V}} \mid i \in I), (P_i \mid i \in I)$ instead of $(\mathcal{V}_i \mid i \in I)$, $\mathcal{S}$, $(s_i \mid i \in I)$, $(\bar{p}_i \mid i \in I)$, respectively.

Proof of the "if" part. We assume that (33.7), (33.8), (33.9) hold. We consider the linear mapping $S : \bigoplus_{i\in I} \mathcal{U}_i \to \mathcal{V}$ defined by (33.1). On account of (33.7) and (33.4) we may define the linear mapping $F : \mathcal{V} \to \bigoplus_{i\in I} \mathcal{U}_i$ by the rule

$$Fv := (E_i v \mid i \in I) \quad \text{for all } v \in \mathcal{V};$$

this mapping is obtained by adjustment from the linear mapping $(E_i \mid i \in I) : \mathcal{V} \to \mathcal{V}^I$. Again by (33.9), we have $SF = 1_{\mathcal{V}}$.

On the other hand, for each $j \in I$, E_j is idempotent by (33.8), and therefore its range $\mathcal{U}_j$ is the set of its fixed points; thus

$$(33.11) \qquad\qquad E_j|_{\mathcal{U}_j} = 1_{\mathcal{U}_j} \quad \text{for all } j \in I.$$

By Proposition 12B and (33.8) and (33.11) we have

$$E_j S u = E_j \sum_I u = E_j \sum_{i\in I} E_i u_i = \sum_{i\in I} E_j E_i u_i = E_j u_j = u_j \quad \text{for all } j \in I \text{ and}$$

$u \in \bigoplus_{i\in I} \mathcal{U}_i$. Therefore

$$FSu = (E_i S u \mid i \in I) = u \quad \text{for all } u \in \bigoplus_{i\in I} \mathcal{U}_i,$$

so that $FS = 1_{\bigoplus_{i\in I} \mathcal{U}_i}$. Thus S is invertible; by Theorem 33A, $(\mathcal{U}_i \mid i \in I)$ is a decomposition of $\mathcal{V}$.

Set $P_i := E_i|^{\mathrm{Rng}} = E_i|^{\mathcal{U}_i}$ for each $i \in I$. By (33.11) and (33.8) we have

$$P_j|_{\mathcal{U}_i} = (E_j 1_{\mathcal{U}_i \subset \mathcal{V}})|^{\mathcal{U}_j} = (E_j E_i)|^{\mathcal{U}_j}_{\mathcal{U}_i} = \begin{cases} E_j|^{\mathcal{U}_j}_{\mathcal{U}_j} = 1_{\mathcal{U}_j} & \text{if } i = j \\ \\ 0_{\mathcal{U}_i \to \mathcal{U}_j} & \text{if } i \neq j. \end{cases} \quad \text{for all } i, j \in I.$$

It follows that $(P_i \mid i \in I)$ is the family of projections of the decomposition $(\mathcal{U}_i \mid i \in I)$. Since $E_i = (E_i|^{\mathrm{Rng}})|^{\mathcal{V}} = P_i|^{\mathcal{V}}$ for every $i \in I$, we conclude that $(E_i \mid i \in I)$ is the family of idempotents of the decomposition. ∎

33G. Corollary. *Let the linear space $\mathcal{V}$ and the decomposition $(\mathcal{U}_i \mid i \in I)$ of $\mathcal{V}$ be given, with the family of projections $(P_i \mid i \in I)$. Then*

$$(33.12) \qquad \mathrm{Supp}(P_i v \mid i \in I) \text{ is finite and } \sum_{i \in I} P_i v = v \text{ for all } v \in \mathcal{V}.$$

Moreover, for every linear space $\mathcal{W}$ and family of linear mappings $(L_i \mid i \in I) \in \underset{i \in I}{\text{\Large$\times$}} \mathrm{Lin}(\mathcal{U}_i, \mathcal{W})$, the unique linear mapping $L : \mathcal{V} \to \mathcal{W}$ satisfying $L|_{\mathcal{U}_i} = L_i$ for all $i \in I$ is given by the rule

$$(33.13) \qquad Lv := \sum_{i \in I} L_i P_i v \quad for \ all \ v \in \mathcal{V}.$$

If $\mathrm{Supp}(\mathcal{U}_i \mid i \in I)$ is finite, (33.13) may be replaced by

$$(33.14) \qquad L := \sum_{i \in I} L_i P_i.$$

Proof. (33.12) follows at once from Proposition 33F ("only if" part) and (33.5). Let $\mathcal{W}$ and $(L_i \mid i \in I)$ be given, and let L be the unique linear mapping from $\mathcal{V}$ to $\mathcal{W}$ satisfying $L|_{\mathcal{U}_i} = L_i$ for all $i \in I$. Then Proposition 12B and (33.12) yield

$$Lv = \sum_{i \in I} L(P_i v) = \sum_{i \in I} (L|_{\mathcal{U}_i}) P_i v = \sum_{i \in I} L_i P_i v \quad \text{for all } v \in \mathcal{V}. \ \blacksquare$$

33H. Corollary. *Let the family of linear spaces $(\mathcal{V}_i \mid i \in I)$ be given. Let $\mathcal{S}$ be the set coproduct-space and $(s_i \mid i \in I)$ the family of insertions of a linear coproduct of $(\mathcal{V}_i \mid i \in I)$. Then $(\mathrm{Rng}s_i \mid i \in I)$ is a decomposition of $\mathcal{S}$.*

Proof. Consider the family of linear mappings $(\bar{p}_i \mid i \in I) \in \underset{i \in I}{\text{\Large$\times$}} \mathrm{Lin}(\mathcal{S}, \mathcal{V}_i)$ defined by (32.4). For each $i \in I$, $\bar{p}_i$ is right-invertible, hence surjective, and therefore

$$(33.15) \qquad \mathrm{Rng}s_i \bar{p}_i = \mathrm{Rng}s_i \quad \text{for all } i \in I.$$

From (32.4) we have

$$(33.16) \qquad s_j \bar{p}_j s_i \bar{p}_i = \begin{cases} s_j \bar{p}_j & \text{if } i = j \\ \\ 0_{\mathcal{S} \to \mathcal{S}} & \text{if } i \neq j \end{cases} \quad \text{for all } i, j \in I.$$

It follows from (33.15), (33.16), (33.11) that (33.7), (33.8), (33.9) hold with $\mathcal{S}$, $(\mathrm{Rng}s_i \mid i \in I)$, $(s_i \bar{p}_i \mid i \in I)$ instead of $\mathcal{V}$, $(\mathcal{U}_i \mid i \in I)$, $(E_i \mid i \in I)$. The assertion then follows by Proposition 33F. $\blacksquare$

We illustrate the use of some of the preceding results.

33I. PROPOSITION. *Let the linear spaces* $\mathcal{V}$ *and* $\mathcal{V}'$, *the decomposition* $(\mathcal{U}_i \mid i \in I)$ *of* $\mathcal{V}$, *the family* $(\mathcal{U}'_i \mid i \in I)$ *of subspaces of* $\mathcal{V}'$, *and the family of linear mappings* $(L_i \mid i \in I) \in \underset{i \in I}{\bigtimes} \mathrm{Lin}(\mathcal{U}_i, \mathcal{U}'_i)$ *be given. Consider the unique linear mapping* $M : \mathcal{V} \to \mathcal{V}'$ *satisfying*

$$(33.17) \qquad\qquad M|_{\mathcal{U}_i} = L_i|^{\mathcal{V}'} \quad \text{for all } i \in I.$$

Of the following statements, any two imply the third:

(i): $(\mathcal{U}'_i \mid i \in I)$ *is a decomposition of* $\mathcal{V}'$.

(ii): L_i *is invertible for every* $i \in I$.

(iii): M *is invertible.*

Proof. We denote by $(P_i \mid i \in I)$ and $(E_i \mid i \in I)$ the families of projections and of idempotents, respectively, of the decomposition $(\mathcal{U}_i \mid i \in I)$ of $\mathcal{V}$.

(i) *and* (ii) *imply* (iii): We consider the unique linear mapping $M' : \mathcal{V}' \to \mathcal{V}$ satisfying

$$(33.18) \qquad\qquad M'|_{\mathcal{U}'_i} = L_i^{-1}|^{\mathcal{V}} \quad \text{for all } i \in I.$$

From (33.1) and (33.18) we have

$$(M'M)|_{\mathcal{U}_i} = M'M1_{\mathcal{U}'_i \subset \mathcal{V}} = M'1_{\mathcal{U}'_i \subset \mathcal{V}'} L_i = 1_{\mathcal{U}_i \subset \mathcal{V}} L_i^{-1} L_i = 1_{\mathcal{U}_i \subset \mathcal{V}} = 1_{\mathcal{V}}|_{\mathcal{U}_i} \quad \text{for all } i \in I.$$

Since $(\mathcal{U}_i \mid i \in I)$ is a decomposition of $\mathcal{V}$, we conclude that $M'M = 1_{\mathcal{V}}$. The proof that $MM' = 1_{\mathcal{V}'}$ is similar; it uses the assumption that $(\mathcal{U}'_i \mid i \in I)$ is a decomposition of $\mathcal{V}'$. Thus M is invertible.

(i) *and* (iii) *imply* (ii). We denote by $(P'_i \mid i \in I)$ the family of projections of the decomposition $(\mathcal{U}'_i \mid i \in I)$.

Let $j \in I$ be given. From (33.17) and (33.4) we have

$$(33.19) \qquad (P_j M^{-1} 1_{\mathcal{U}'_j \subset \mathcal{V}'}) L_j = P_j M^{-1} M 1_{\mathcal{U}_j \subset \mathcal{V}} = P_j 1_{\mathcal{U}_j \subset \mathcal{V}} = 1_{\mathcal{U}_j}.$$

On the other hand, from (33.17), (33.4), (33.5) we have

$$P'_j M E_i = P'_j M 1_{\mathcal{U}_i \subset \mathcal{V}} P_i = P'_j 1_{\mathcal{U}'_i \subset \mathcal{V}_i} L_i P_i = \begin{cases} L_j P_j & \text{if } i = j \\ \\ 0_{\mathcal{V} \to \mathcal{U}'_j} & \text{if } i \in I \backslash \{j\}. \end{cases}$$

By Propositions 33F and 12B we then have

$$P_j' M v = P_j' M \sum_{i \in I} E_i v = \sum_{i \in I} P_j' M E_i v = L_j P_j v \quad \text{for all } v \in \mathcal{V},$$

so that $P_j' M = L_j P_j$. It follows that

$$(33.20) \qquad L_j(P_j M^{-1} 1_{\mathcal{U}_j' \subset \mathcal{V}'}) = P_j' M M^{-1} 1_{\mathcal{U}_j' \subset \mathcal{V}'} = P_j' 1_{\mathcal{U}_j' \subset \mathcal{V}'} = 1_{\mathcal{U}_j'}.$$

From (33.19) and (33.20) it follows that L_j is invertible. It remains to observe that $j \in I$ was arbitrary.

(ii) *and* (iii) *imply* (i). We define the family $(E_i' \mid i \in I) \in (\mathrm{Lin} \mathcal{V}')^I$ by the rule

$$E_i' := M E_i M^{-1} \quad \text{for all } i \in I.$$

Since L_i is invertible, and hence surjective, for every $i \in I$, we have, by (33.17) and Proposition 33F,

$$\mathrm{Rng} E_i' = M_{>}(\mathrm{Rng} E_i) = M_{>}(\mathcal{U}_i) = \mathrm{Rng}(M|_{\mathcal{U}_i}) = \mathrm{Rng}\,(L_i|^{\mathcal{V}'}) = \mathrm{Rng} L_i = \mathcal{U}_i' \quad \text{for all } i \in I,$$

$$E_j' E_i' = M E_j M^{-1} M E_i M^{-1} = M E_j E_i M^{-1} = \delta_{i,j}^I M E_i M^{-1} = \delta_{i,j}^I E_i' \quad \text{for all } i, j \in I.$$

Moreover, for every $v' \in \mathcal{V}'$, $\mathrm{Supp}(E_i' v' \mid i \in I) = \mathrm{Supp}(E_i M^{-1} v' \mid i \in I)$ is finite, and, using Proposition 12B,

$$\sum_{i \in I} E_i' v' = \sum_{i \in I} M E_i M^{-1} v' = M \sum_{i \in I} E_i M^{-1} v' = M M^{-1} v' = v'.$$

By Proposition 33F it follows that $(\mathcal{U}_i' \mid i \in I)$ is a decomposition of $\mathcal{V}'$.

We observe that we have not used explicitly the assumption that L_i is injective for each $i \in I$; but this assumption is redundant, since it follows from (33.17) and the assumed injectivity of M. ∎

Chapter 4

FAMILIES IN LINEAR SPACES

41. Linear combination

Let the linear space $\mathcal{V}$ and the family $z \in \mathcal{V}^I$ be given. The mapping $\mathrm{lc}_z^{\mathcal{V}} : \mathbb{F}^{(I)} \to \mathcal{V}$ defined by the rule

$$(41.1) \qquad \mathrm{lc}_z^{\mathcal{V}} a := \sum_{i \in I} a_i z_i \quad \text{for all } a \in \mathbb{F}^{(I)}$$

is called the **linear-combination mapping for** z **(in $\mathcal{V}$)**. Its value $\mathrm{lc}_z^{\mathcal{V}} a$ at $a \in \mathbb{F}^{(I)}$ is called the **linear combination of** z **(in $\mathcal{V}$) with the coefficient family** a. Evaluation of $\mathrm{lc}_z^{\mathcal{V}}$ at the Kronecker families $\delta_j^I \in \mathbb{F}^{(I)}$ for all $j \in I$ yields

$$(41.2) \qquad \mathrm{lc}_z^{\mathcal{V}} \delta_j^I = \sum_{i \in I} \delta_{j,i}^I z_i = 1 z_j = z_j \quad \text{for all } j \in I.$$

41A. Proposition. *Let the linear space $\mathcal{V}$ and the family $z \in \mathcal{V}^I$ be given. Then $\mathrm{lc}_z^{\mathcal{V}}$ is the only solution of the problem*

$$(41.3) \qquad ?L \in \mathrm{Lin}\,(\mathbb{F}^{(I)}, \mathcal{V}), \ (\forall i \in I, \ L\delta_i^I = z_i).$$

In particular, $\mathrm{lc}_z^{\mathcal{V}}$ is linear.

Proof. Let L be a solution of (41.3). Let $a \in \mathbb{F}^{(I)}$ be given. By (13.4) and Proposition 12B,

$$La = L \sum_{i \in I} a_i \delta_i^I = \sum_{i \in I} L(a_i \delta_i^I) = \sum_{i \in I} a_i L \delta_i^I = \sum_{i \in I} a_i z_i = \mathrm{lc}_z^{\mathcal{V}} a.$$

Since $a \in \mathbb{F}^{(I)}$ was arbitrary, it follows that $L = \mathrm{lc}_z^{\mathcal{V}}$.

Conversely, $\mathrm{lc}_z^{\mathcal{V}}$ is the composite of the mapping $(a \mapsto (a_i z_i \mid i \in I)) : \mathbb{F}^{(I)} \to \mathcal{V}^{(I)}$, which is obviously linear, and the linear mapping $(v \mapsto \sum_I v) : \mathcal{V}^{(I)} \to \mathcal{V}$ (Example 13G,(e)). Therefore $\mathrm{lc}_z^{\mathcal{V}} \in \mathrm{Lin}(\mathbb{F}^{(I)}, \mathcal{V})$. By (41.2) $\mathrm{lc}_z^{\mathcal{V}}$ is then a solution of (41.3).

An alternative proof may be obtained from Proposition 32B. ∎

41B. Remarks. (a): Let the linear space $\mathcal{V}$ and the subspace $\mathcal{U}$ of $\mathcal{V}$ be given. It follows from Remark 13E,(d) that for every family $z \in \mathcal{U}^I$ we have

$$\mathrm{lc}_z^{\mathcal{V}} = \mathrm{lc}_z^{\mathcal{U}}|^{\mathcal{V}}.$$

We may therefore write $1c_z a$ for every $a \in \mathbb{F}^{(I)}$, omitting the indication of the space without causing ambiguity.

(b): Let the linear space $\mathcal{V}$ and the subset $\mathcal{A}$ of $\mathcal{V}$ be given. In accordance with the notational convention established in *Basic Language*, Section 42, $\mathrm{lc}_{\mathcal{A}}^{\mathcal{V}}$ denotes the linear-combination mapping for the set $\mathcal{A}$ self-indexed as a family. Thus

$$\mathrm{lc}_{\mathcal{A}}^{\mathcal{V}} a = \sum_{v \in \mathcal{A}} a_v v \quad \text{for all } a \in \mathbb{F}^{(\mathcal{A})}. \quad ∎$$

Linear combinations can be used to describe the linear spans of sets: specifically, we shall show that the linear span of a subset of a linear space is precisely the set of all linear combinations of that subset (self-indexed).

41C. Theorem. *Let the linear space $\mathcal{V}$ and the family $z \in \mathcal{V}^I$ be given. Then*

$$\mathrm{Lsp\ Rng}\ z = \mathrm{Rng}\ \mathrm{lc}_z^{\mathcal{V}}.$$

Proof. Set $\mathcal{U} := \mathrm{Lsp\ Rng}\ z$; this is a subspace of $\mathcal{V}$, and we have $z \in \mathcal{U}^I$. By Remark 41B,(a), $\mathrm{Rng}\ \mathrm{lc}_z^{\mathcal{V}} = \mathrm{Rng}\ \mathrm{lc}_z^{\mathcal{U}} \subset \mathcal{U} = \mathrm{Lsp\ Rng}\ z$. On the other hand, (41.2) shows that $\mathrm{Rng}\,z \subset \mathrm{Rng}\ \mathrm{lc}_z^{\mathcal{V}}$. But $\mathrm{Rng}\ \mathrm{lc}_z^{\mathcal{V}}$ is a subspace of $\mathcal{V}$, since $\mathrm{lc}_z^{\mathcal{V}}$ is linear. Therefore $\mathrm{Lsp\ Rng}\ z \subset \mathrm{Rng}\ \mathrm{lc}_z^{\mathcal{V}}$. ∎

41D. Corollary. *Let the linear space $\mathcal{V}$ and the subset $\mathcal{A}$ of $\mathcal{V}$ be given. Then*

$$\mathrm{Lsp}\mathcal{A} = \mathrm{Rng}\ \mathrm{lc}_{\mathcal{A}}^{\mathcal{V}} = \{\sum_{v \in \mathcal{A}} a_v v \mid a \in \mathbb{F}^{(\mathcal{A})}\}.$$

We can characterize the linearity of mappings by means of linear combinations: a mapping from a linear space to a linear space is linear if and only if it "preserves linear combinations", as we now show.

41E. Theorem. *Let the linear spaces $\mathcal{V}$ and $\mathcal{W}$ and the mapping $L : \mathcal{V} \to \mathcal{W}$ be given. Then L is linear if and only if*

$$(41.4) \qquad L \circ \mathrm{lc}_z^{\mathcal{V}} = \mathrm{lc}_{Loz}^{\mathcal{W}} \quad \text{for all sets } I \text{ and families } z \in \mathcal{V}^I.$$

Proof. Assume that L is linear. For every family $z \in \mathcal{V}^I$ we have, by Proposition 12B,

$$L\mathrm{lc}_z^{\mathcal{V}} a = L \sum_{i \in I} a_i z_i = \sum_{i \in I} L(a_i z_i) = \sum_{i \in I} a_i L z_i = \mathrm{lc}_{Lz}^{\mathcal{W}} a \quad \text{for all } a \in \mathbb{F}^{(I)}.$$

Therefore (41.4) holds. Assume, conversely, that (41.4) holds. For all $u, v \in \mathcal{V}$ and $s \in \mathbb{F}$, we apply (41.4) to the lists (u, v) and (u), of length 2 and 1, respectively, and find

$$
\begin{aligned}
L(u + v) \;&=\; L(1u + 1v) = L(\mathrm{lc}_{(u,v)}(1, 1)) = \mathrm{lc}_{(L(u), L(v))}(1, 1) = \\
&=\; 1L(u) + 1L(v) = L(u) + L(v) \\
L(su) \;&=\; L(\mathrm{lc}_{(u)}(s)) = \mathrm{lc}_{(L(u))}(s) = sL(u);
\end{aligned}
$$

we conclude that L is linear. ∎

41F. PROPOSITION. *Let the linear space $\mathcal{V}$ and the mapping $\Omega : I \to K$ be given. For all families $z \in \mathcal{V}^K$ and $a \in \mathbb{F}^{(I)}$ we have*

$$
\Big(\sum_{\Omega^{<}(\{k\})} a \mid k \in K \Big) \in \mathbb{F}^{(K)} \quad \text{and} \quad \mathrm{lc}_{z \circ \Omega}\, a = \mathrm{lc}_z \Big(\sum_{\Omega^{<}(\{k\})} a \mid k \in K \Big).
$$

Proof. Let $z \in \mathcal{V}^K$ and $a \in \mathbb{F}^{(I)}$ be given. By *Basic Language*, Theorem 114A we find that $\mathrm{Supp}\big(\sum_{\Omega^{<}(\{k\})} a \mid k \in K \big)$ is indeed finite, and that

$$
\begin{aligned}
\mathrm{lc}_{z \circ \Omega}\, a &= \sum_{i \in I} a_i z_{\Omega(i)} = \sum_{k \in K} \sum_{i \in \Omega^{<}(\{k\})} a_i z_{\Omega(i)} = \sum_{k \in K} \sum_{i \in \Omega^{<}(\{k\})} a_i z_k = \\
&= \sum_{k \in K} \Big(\sum_{\Omega^{<}(\{k\})} a \Big) z_k = \mathrm{lc}_z \Big(\sum_{\Omega^{<}(\{k\})} a \mid k \in K \Big). \;\blacksquare
\end{aligned}
$$

41G. COROLLARY. *Let the linear space $\mathcal{V}$ and the bijection $\Omega : I \to K$ be given. Then*

$$
\mathrm{lc}_{z \circ \Omega}\, a = \mathrm{lc}_z(a \circ \Omega^{\leftarrow}) \quad \text{for all } a \in \mathbb{F}^{(I)} \text{ and } z \in \mathcal{V}^K.
$$

42. Linear independence

Let the linear space $\mathcal{V}$ be given. A family $z \in \mathcal{V}^I$ is said to be **linearly indepen-dent (in** $\mathcal{V}$**)** if $\mathrm{lc}_z^{\mathcal{V}}$ is injective, and **linearly dependent (in** $\mathcal{V}$**)** otherwise. A subset $\mathcal{A}$ of $\mathcal{V}$ is said to be **linearly independent (in** $\mathcal{V}$**)** if $\mathcal{A}$ self-indexed is linearly inde-pendent (in $\mathcal{V}$), i.e., if $\mathrm{lc}_{\mathcal{A}}^{\mathcal{V}}$ is injective; otherwise, $\mathcal{A}$ is said to be **linearly dependent (in** $\mathcal{V}$**)**.

42A. REMARKS. (a): Let the subspace $\mathcal{U}$ of $\mathcal{V}$ be given. By Remark 41B,(a), a family $z \in \mathcal{U}^I$ is linearly independent in $\mathcal{U}$ if and only if z is linearly independent in $\mathcal{V}$. The phrases "linearly independent" and "linearly dependent" are therefore unambiguous without the indication of the space. The same comment applies to the linear independence or dependence of subsets of $\mathcal{U}$.

(b): By Corollary 14D, a family $z \in \mathcal{V}^I$ is linearly independent if and only if Null $\mathrm{lc}_z^{\mathcal{V}} = \{0\}$, i.e.,

$$\forall a \in \mathbb{F}^{(I)}, \quad \sum_{i \in I} a_i z_i = 0 \quad \Rightarrow \quad a = 0.$$

(c): Let the linearly independent family $z \in \mathcal{V}^I$ be given. By (b) and (1.2), $z_i = \mathrm{lc}_z \delta_i^I \neq 0$ for all $i \in I$, so that $0 \notin \mathrm{Rng}z$. In particular, no linearly independent subset of $\mathcal{V}$ contains 0. ∎

42B. EXAMPLE. Let the linear space $\mathcal{V}$ be given. The empty family is linearly independent in $\mathcal{V}$. A single-index family in $\mathcal{V}$ is linearly independent if and only if the single term is not 0. For all $u, v \in \mathcal{V}$, the list (u, v) is linearly independent if and only if $u \neq 0$ and $v \notin \mathbb{F}u$. A characterization of linearly independent lists of all lengths will be given in Corollary 42L. ∎

The next result shows that some questions regarding the linear independence of families may be reduced to questions regarding the linear independence of sets.

42C. PROPOSITION. *Let the linear space $\mathcal{V}$ be given. A family $z \in \mathcal{V}^I$ is linearly independent if and only if z is injective and $\mathrm{Rng}z$ is linearly independent.*

Proof. Assume that $z \in \mathcal{V}^I$ is linearly independent. From (41.2) we obtain the following chain of implications for all $i, j \in I$:

$$z_i = z_j \quad \Rightarrow \quad \mathrm{lc}_z \delta_i^I = \mathrm{lc}_z \delta_j^I \quad \Rightarrow \quad \delta_i^I = \delta_j^I \quad \Rightarrow \quad \delta_{i,j}^I = \delta_{j,j}^I = 1 \quad \Rightarrow \quad i = j.$$

This shows that z is injective.

Now assume that z is injective, and define the bijection $\Omega := z|^{\mathrm{Rng}z}$ (Ω is actually the family z itself, regarded as a surjective mapping.) Then $z = (\mathrm{Rng}z) \circ \Omega$, where $\mathrm{Rng}z$ is self-indexed. By Corollary 41G, $\mathrm{lc}_z^{\mathcal{V}} = \mathrm{lc}_{(\mathrm{Rng}z) \circ \Omega}^{\mathcal{V}}$ is the composite of the linear isomorphism $(a \mapsto a \circ \Omega^{\leftarrow}) : \mathbb{F}^{(I)} \to \mathbb{F}^{(\mathrm{Rng}z)}$ and $\mathrm{lc}_{\mathrm{Rng}z}^{\mathcal{V}}$. Therefore $\mathrm{lc}_z^{\mathcal{V}}$ is injective if and only if $\mathrm{lc}_{\mathrm{Rng}z}^{\mathcal{V}}$ is injective; in other words, z is linearly independent if and only if $\mathrm{Rng}z$ is linearly independent. ∎

42D. PROPOSITION. *Let the linear space $\mathcal{V}$ be given.*

(a): *Every restriction of a linearly independent family in $\mathcal{V}$ is linearly independent.*

(b): *A family in $\mathcal{V}$ is linearly independent if (and only if) its restriction to each finite subset of the index set is linearly independent.*

Proof. Proof of (a). Let the linearly independent family $z \in \mathcal{V}^I$ and the subset K of I be given. Let $b \in$ Null $\mathrm{lc}_{z|_K}^{\mathcal{V}}$ be given, and define $a \in \mathbb{F}^{(I)}$ by requiring $a|_K := b$ and $a|_{I \setminus K} := 0$. Then

$$\mathrm{lc}_z a = \mathrm{lc}_{z|_K}(a|_K) + \mathrm{lc}_{z|_{I \setminus K}}(a|_{I \setminus K}) = \mathrm{lc}_{z|_K} b = 0,$$

and hence, by Remark 42A,(b), $a \in$ Null $\mathrm{lc}_z^{\mathcal{V}} = \{0\}$. Therefore $a = 0$, and $b = a|_K = 0$. Since $b \in$ Null $\mathrm{lc}_{z|_K}^{\mathcal{V}}$ was arbitrary, we have Null $\mathrm{lc}_{z|_K}^{\mathcal{V}} = \{0\}$, and $z|_K$ is linearly independent.

Proof of (b). Let the linearly dependent family $z \in \mathcal{V}^I$ be given. We may choose $a \in ($Null $\mathrm{lc}_z^{\mathcal{V}})^{\times}$ by Remark 42A,(b). Then $K := \mathrm{Supp}\, a$ is a finite non-empty subset of I, and we have $a|_{I \setminus K} = 0$, $a|_K \neq 0$. Therefore

$$\mathrm{lc}_{z|_K}(a|_K) = \mathrm{lc}_{z|_K}(a|_K) + \mathrm{lc}_{z|_{I \setminus K}}(a|_{I \setminus K}) = \mathrm{lc}_z a = 0,$$

and so $a|_K \in ($Null $\mathrm{lc}_{z|_K}^{\mathcal{V}})^{\times}$. Therefore $z|_K$ is linearly dependent, and K is a finite subset of I. ∎

42E. COROLLARY. *Let the linear space $\mathcal{V}$ be given.*

(a): *Every subset of a linearly independent subset of $\mathcal{V}$ is linearly independent.*

(b): *A subset of $\mathcal{V}$ is linearly independent if (and only if) each of its finite subsets is linearly independent. In other words, the collection $\{\mathcal{A} \in \mathfrak{P}(\mathcal{V}) \mid \mathcal{A}$ is linearly independent$\}$ is of finitary character.*

We present some characterizations of linear independence.

42F. LEMMA. *Let the linear space $\mathcal{V}$, the family $z \in \mathcal{V}^I$, and the index $j \in I$ be given. The following statements are equivalent.*

(i): *There exists $a \in$ Null $\mathrm{lc}_z^{\mathcal{V}}$ with $a_j \neq 0$.*

(ii): *$z_j \in$ Lsp $\mathrm{Rng}(z|_{I \setminus \{j\}})$.*

(iii): *Lsp $\mathrm{Rng}(z|_{I \setminus \{j\}}) =$ Lsp $\mathrm{Rng}\, z$.*

Proof. (i) *implies* (ii). Choose $a \in$ Null $\mathrm{lc}_z^{\mathcal{V}}$ with $a_j \neq 0$. Then

$$0 = \mathrm{lc}_z a = a_j z_j + \mathrm{lc}_{z|_{I \setminus \{j\}}}(a|_{I \setminus \{j\}})$$

and therefore, by Theorem 41C,

$$z_j = -\frac{1}{a_j} \mathrm{lc}_{z|_{I \setminus \{j\}}}(a|_{I \setminus \{j\}}) \in -\frac{1}{a_j} \mathrm{Rng}\, \mathrm{lc}_{z|_{I \setminus \{j\}}}^{\mathcal{V}} = \mathrm{Lsp}\, \mathrm{Rng}(z|_{I \setminus \{j\}}).$$

(ii) *implies* (i). By Theorem 41C, $z_j \in$ Lsp $\mathrm{Rng}(z|_{I \setminus \{j\}}) = \mathrm{Rng}\, \mathrm{lc}_{z|_{I \setminus \{j\}}}^{\mathcal{V}}$. We may therefore choose $b \in \mathbb{F}^{(I \setminus \{j\})}$ such that $z_j = \mathrm{lc}_{z|_{I \setminus \{j\}}}^{\mathcal{V}} b$. We define $a \in \mathbb{F}^{(I)}$ by requiring $a_j := -1$ and $a|_{I \setminus \{j\}} = b$. Then $a_j \neq 0$ and

$$\mathrm{lc}_z a = a_j z_j + \mathrm{lc}_{z|_{I\setminus\{j\}}}(a|_{I\setminus\{j\}}) = -z_j + \mathrm{lc}_{z|_{I\setminus\{j\}}} b = 0,$$

so that $a \in \mathrm{Null}\, \mathrm{lc}_z^{\mathcal{V}}$.

(ii) *is equivalent to* (iii). By Lemma 12H,

$$\mathrm{Lsp}\, \mathrm{Rng} z = \mathrm{Lsp}(\mathrm{Rng}(z|_{I\setminus\{j\}}) \cup \{z_j\}) = \mathrm{Lsp}\, \mathrm{Rng}(z|_{I\setminus\{J\}}) + \mathbb{F} z_j.$$

Therefore

$$z_j \in \mathrm{Lsp}\, \mathrm{Rng}(z|_{I\setminus\{j\}}) \ \Leftrightarrow\ \mathbb{F} z_j \subset \mathrm{Lsp}\, \mathrm{Rng}(z|_{I\setminus\{j\}}) \ \Leftrightarrow\ \mathrm{Lsp}\, \mathrm{Rng}\, z \subset \mathrm{Lsp}\, \mathrm{Rng}(z|_{I\setminus\{j\}})$$

but obviously $\mathrm{Lsp}\, \mathrm{Rng}(z|_{I\setminus\{j\}}) \subset \mathrm{Lsp}\, \mathrm{Rng}\, z$. ∎

42G. Theorem. *Let the linear space $\mathcal{V}$, the family $z \in \mathcal{V}^I$, and the subset K of I be given. The following statements are equivalent.*

(i): *z is linearly independent.*

(ii): *$z|_K$ is linearly independent, and $z_j \notin \mathrm{Lsp}\, \mathrm{Rng}(z|_{I\setminus\{j\}})$ for all $j \in I\setminus K$.*

(iii): *$z|_K$ is linearly independent, and $\mathrm{Lsp}\, \mathrm{Rng}(z|_{I\setminus\{j\}}) \subsetneqq \mathrm{Lsp}\, \mathrm{Rng}\, z$ for all $j \in I\setminus K$.*

Proof. (i) *implies* (ii). This follows at once from Proposition 42D,(a) and from lemma 42F,((ii) $\Rightarrow$ (i)) by contraposition.

(ii) *implies* (i). Let $a \in \mathrm{Null}\, \mathrm{lc}_z^{\mathcal{V}}$ be given. Since $z_j \notin \mathrm{Lsp}\, \mathrm{Rng}(z|_{I\setminus\{j\}})$ for all $j \in I\setminus K$, it follows by contraposition from Lemma 42F,((i) $\Rightarrow$ (ii)) that $a|_{I\setminus K} = 0$. Then

$$0 = \mathrm{lc}_z a = \mathrm{lc}_{z|_K}(a|_K) + \mathrm{lc}_{z|_{I\setminus K}}(a|_{I\setminus K}) = \mathrm{lc}_{z|_K}(a|_K).$$

Then $a|_K \in \mathrm{Null}\, \mathrm{lc}_{z|_K}^{\mathcal{V}} = \{0\}$, since $z|_K$ is linearly independent. Thus $a|_K = 0$ and $a|_{I\setminus K} = 0$; therefore $a = 0$. Since $a \in \mathrm{Null}\, \mathrm{lc}_z^{\mathcal{V}}$ was arbitrary, we have $\mathrm{Null}\, \mathrm{lc}_z^{\mathcal{V}} = \{0\}$, and z is linearly independent.

(ii) *is equivalent to* (iii). This is immediate, by contraposition, from Lemma 42F,((ii) $\Leftrightarrow$ (iii)). ∎

It is useful to record two extreme cases of Theorem 42G.

42H. Corollary. *Let the linear space $\mathcal{V}$ and the family $z \in \mathcal{V}^I$ be given. The following statements are equivalent:*

(i): *z is linearly independent.*

(ii): *$z_j \notin \mathrm{Lsp}\, \mathrm{Rng}(z|_{I\setminus\{j\}})$ for all $j \in I$.*

(iii): *$\mathrm{Lsp}\, \mathrm{Rng}(z|_{I\setminus\{j\}}) \subsetneqq \mathrm{Lsp}\, \mathrm{Rng}\, z$ for all $j \in I$.*

(iv): *$0 \notin \mathrm{Rng} z$, and the family $(\mathbb{F} z_i \mid i \in I)$ of subspaces of $\mathcal{V}$ is disjunct.*

Proof. The equivalence of (i), (ii), (iii) follows from Theorem 42G with $K := \emptyset$.

(ii) *is equivalent to* (iv). For every $j \in I$ we have, by (13.13), $\mathrm{Lsp}\, \mathrm{Rng}(z|_{I\setminus\{j\}}) =$

$$= \sum_{i\in I\setminus\{j\}} \mathbb{F} z_i;\ \text{therefore}$$

$$z_j \notin \mathrm{Lsp} \; \mathrm{Rng}(z|_{I \setminus \{j\}}) \quad \Leftrightarrow \quad z_j \notin \sum_{i \in I \setminus \{j\}} \mathbb{F} z_i \quad \Leftrightarrow \quad \mathbb{F}^\times z_j \cap \sum_{i \in I \setminus \{j\}} \mathbb{F} z_i = \varnothing \quad \Leftrightarrow$$

$$\Leftrightarrow \quad (z_j \neq 0 \quad \text{and} \quad \mathbb{F} z_j \cap \sum_{i \in I \setminus \{j\}} \mathbb{F} z_i = \{0\}).$$

The validity for all $j \in I$ of the right-most assertion in this chain of equivalences is precisely the content of (iv) (cf. Remark 33D,(b)). ∎

42I. COROLLARY. *Let the linear space $\mathcal{V}$, the family $z \in \mathcal{V}^I$, and the index $j \in I$ be given. The following statements are equivalent.*

(i): *z is linearly independent.*

(ii): *$z|_{I \setminus \{j\}}$ is linearly independent, and $z_j \notin \mathrm{Lsp} \; \mathrm{Rng}(z|_{I \setminus \{j\}})$.*

(iii): *$z|_{I \setminus \{j\}}$ is linearly independent, and $\mathrm{Lsp} \; \mathrm{Rng}(z|_{I \setminus \{j\}}) \subsetneqq \mathrm{Lsp} \; \mathrm{Rng} z$.*

We next record a version of Theorem 42G for subsets of a linear space.

42J. COROLLARY. *Let the linear space $\mathcal{V}$, the subset $\mathcal{A}$ of $\mathcal{V}$, and the subset $\mathcal{B}$ of $\mathcal{A}$ be given. The following statements are equivalent.*

(i): *$\mathcal{A}$ is linearly independent.*

(ii): *$\mathcal{B}$ is linearly independent, and $v \notin \mathrm{Lsp}(\mathcal{A} \setminus \{v\})$ for all $v \in \mathcal{A} \setminus \mathcal{B}$.*

(iii): *$\mathcal{B}$ is linearly independent, and $\mathrm{Lsp}(\mathcal{A} \setminus \{v\}) \subsetneqq \mathrm{Lsp} \mathcal{A}$ for all $v \in \mathcal{A} \setminus \mathcal{B}$.*

(iv): *$\mathcal{B}$ is linearly independent, and $\{\mathcal{S} \in \mathfrak{P}(\mathcal{A}) \mid \mathcal{B} \subset \mathcal{S}, \; \mathrm{Lsp} \mathcal{S} = \mathrm{Lsp} \mathcal{A}\} = \{\mathcal{A}\}$.*

Proof. The equivalence of (i), (ii), (iii) follows from Theorem 42G. Set $\Gamma := \{\mathcal{S} \in \mathfrak{P}(\mathcal{A}) \mid \mathcal{B} \subset \mathcal{S}, \; \mathrm{Lsp} \mathcal{S} = \mathrm{Lsp} \mathcal{A}\}$. We certainly have $\mathcal{A} \in \Gamma$.

(iii) *implies* (iv). Let $\mathcal{S} \in \Gamma$ be given. For every $v \in \mathcal{A} \setminus \mathcal{S}$ we should have $v \in \mathcal{A} \setminus \mathcal{B}$ and $\mathcal{S} \subset \mathcal{A} \setminus \{v\}$; and therefore, by (iii),

$$\mathrm{Lsp} \mathcal{S} \subset \mathrm{Lsp}(\mathcal{A} \setminus \{v\}) \subsetneqq \mathrm{Lsp} \mathcal{A} = \mathrm{Lsp} \mathcal{S} \quad \text{for all } v \in \mathcal{A} \setminus \mathcal{S},$$

which cannot be unless $\mathcal{A} \setminus \mathcal{V} = \varnothing$, i.e., $\mathcal{S} = \mathcal{A}$. Hence $\Gamma = \{\mathcal{A}\}$.

(iv) *implies* (iii). Let $v \in \mathcal{A} \setminus \mathcal{B}$ be given. Since $\mathcal{A} \setminus \{v\} \neq \mathcal{A}$, we have $\mathcal{A} \setminus \{v\} \notin \Gamma$, but $\mathcal{B} \subset \mathcal{A} \setminus \{v\}$. Therefore $\mathrm{Lsp}(\mathcal{A} \setminus \{v\}) \subsetneqq \mathrm{Lsp} \mathcal{A}$. ∎

In the following proposition we have an ordered set I. We recall the notation $\mathrm{Spr}(j) := \mathrm{Lb}(\{j\}) \setminus \{j\}$ for the set of all members of I that strictly precede j, for each $j \in I$.

42K. PROPOSITION. *Let the linear space $\mathcal{V}$ and the totally ordered set I be given. A family $z \in \mathcal{V}^I$ is linearly independent if and only if*

$$(42.1) \qquad\qquad z_j \notin \mathrm{Lsp} \; \mathrm{Rng}(z|_{\mathrm{Spr}(j)}) \quad \text{for all } j \in I.$$

Proof. Proof of the "only if" part. Assume that z is linearly independent. By Corollary 42H we have

$$\mathrm{Lsp} \; \mathrm{Rng}(z|_{\mathrm{Spr}(j)}) \subset \mathrm{Lsp} \; \mathrm{Rng}(z|_{I \setminus \{j\}}) \subset \mathcal{V} \setminus \{z_j\} \quad \text{for all } j \in I,$$

so that (42.1) holds.

Proof of the "if" part. We assume that (42.1) holds. We shall prove by general induction that

$$P(S) :\Leftrightarrow z|_S \text{ is linearly independent}$$

holds for all $S \in \mathfrak{F}(I)$. It will then follow by Proposition 42D,(b) that z itself is linearly independent, and the proof will be complete.

Let then $S \in \mathfrak{F}(I)$ be given, such that $P(T)$ holds for all proper subsets T of S. If $S = \emptyset$, then $z|_S$ is empty, and therefore linearly independent, so that $P(S)$ holds in this case. Suppose now that $S \neq \emptyset$. By *Basic Language*, Corollary 105B, the totally ordered non-empty finite ordered subset S of I has a maximum. By the induction hypothesis, $P(S\backslash\{\max S\})$ holds, i.e.,

$$(42.2) \qquad z|_{S\backslash\{\max S\}} \text{ is linearly independent.}$$

By (42.1) we have

$$\text{Lsp Rng}(z|_{S\backslash\{\max S\}}) \subset \text{Lsp Rng}(z|_{\text{Spr}(\max S)}) \subset \mathcal{V}\backslash\{z_{\max S}\},$$

and therefore

$$(42.3) \qquad z_{\max S} \notin \text{Lsp Rng}(z|_{S\backslash\{\max S\}}).$$

By Corollary 42I it follows from (42.2) and (42.3) that $z|_S$ is linearly independent, so that $P(S)$ holds in this case too. This completes the induction step. ∎

42L. COROLLARY. *Let the linear space $\mathcal{V}$ be given. A list $(z_k \mid k \in n^\sqsubset)$ [a sequence $(z_k \mid k \in \mathbb{N})$] in $\mathcal{V}$ is linearly independent if and only if*

$$z_m \neq \text{Lsp}\{z_k \mid k \in m^\sqsubset\} \quad \text{for all } m \in n^\sqsubset \text{ [all } m \in \mathbb{N}].$$

More results concerning linear independence of families and sets will be presented in the next section.

43. Linear independence, spanning, bases

Let the linear space $\mathcal{V}$ be given. A family $z \in \mathcal{V}^I$ is said to **span** $\mathcal{V}$, or to be **spanning in** $\mathcal{V}$ if Lsp Rng $z = \mathcal{V}$; it is called a **basis of** $\mathcal{V}$ if z is both linearly independent and spanning in $\mathcal{V}$. A subset $\mathcal{A}$ of $\mathcal{V}$ is said to **span** $\mathcal{V}$ or to be **spanning in** $\mathcal{V}$ if Lsp$\mathcal{A} = \mathcal{V}$, i.e., if $\mathcal{A}$ self-indexed spans $\mathcal{V}$; $\mathcal{A}$ is called a **basis (-set)** of $\mathcal{V}$ if $\mathcal{A}$ is both linearly independent and spanning in $\mathcal{V}$, i.e., if $\mathcal{A}$ self-indexed is a basis of $\mathcal{V}$.

43A. REMARK. A family $z \in \mathcal{V}^I$ spans $\mathcal{V}$ if (and only if) some restriction of z spans $\mathcal{V}$. A subset $\mathcal{A}$ of $\mathcal{V}$ spans $\mathcal{V}$ if (and only if) some subset of $\mathcal{A}$ spans $\mathcal{V}$. ∎

43B. PROPOSITION. *Let the linear space $\mathcal{V}$ and the family $z \in \mathcal{V}^I$ be given.*

(a): *z spans $\mathcal{V}$ if and only if $\mathrm{lc}_z^{\mathcal{V}}$ is surjective; z is a basis of $\mathcal{V}$ if and only if $\mathrm{lc}_z^{\mathcal{V}}$ is invertible.*

(b): *z spans $\mathcal{V}$ if and only if $\mathrm{Rng}z$ spans $\mathcal{V}$; z is a basis of $\mathcal{V}$ if and only if z is injective and $\mathrm{Rng}z$ is a basis-set of $\mathcal{V}$.*

(c): *z is linearly independent if and only if z is a basis of Lsp Rng z.*

(d): *z is a basis of $\mathcal{V}$ if and only if $0 \notin \mathrm{Rng}z$ and the family $(\mathbb{F}z_i \mid i \in I)$ is a decomposition of $\mathcal{V}$.*

Proof. (a) follows from Theorem 41C and the definitions. (b) follows from the definitions and Proposition 42C. (c) follows from the definitions. According to Corollary 42H,((i) ⟺ (iv)), z is a basis of $\mathcal{V}$ if and only if $0 \notin \mathrm{Rng}z$, $(\mathbb{F}z_i \mid i \in I)$ is disjunct, and Lsp Rng $z = \mathcal{V}$; but Lsp Rng $z = \sum_{i \in I} \mathbb{F}z_i$, by (13.13); (d) then follows from Theorem 33A, ((i) ⟺ (iii)). ∎

43C. EXAMPLES. (a): The empty set is a basis-set of the linear space $\mathcal{V}$ if and only if $\mathcal{V}$ is a zero-space; it is then the only basis-set of $\mathcal{V}$. The basis-sets of the linear space $\mathbb{F}$ are precisely the singletons $\{t\}$ with $t \in \mathbb{F}^{\times}$. The singleton $\{1\}$ may be regarded as the "natural" basis-set of $\mathbb{F}$.

(b): Let the set I be given. We consider the family $\delta^I : (\delta_i^I \mid i \in I)$ in $\mathbb{F}^{(I)}$. (This is the family of rows of the Kronecker matrix δ^I– with entries the zero and unity of $\mathbb{F}$ – and we use the same symbol δ^I since there is no danger of confusion and the two objects correspond in the obvious natural isomorphism from $\mathbb{F}^{I \times I}$ to $(\mathbb{F}^I)^I$.) By Proposition 41A, $1c_{\delta^I}^{\mathbb{F}^{(I)}}$ is the only solution of

$$?L \in \mathrm{Lin}(\mathbb{F}^{(I)}, \mathbb{F}^{(I)}), \ L\delta^I = \delta^I.$$

But $1_{\mathbb{F}^{(I)}}$ is of course a solution too. Therefore $1c_{\delta^I}^{\mathbb{F}^{(I)}} = 1_{\mathbb{F}^{(I)}}$, and therefore δ^I is a basis of $\mathbb{F}^{(I)}$ (Proposition 43B,(a)); it is called the **Kronecker basis**, or **natural basis**, of $\mathbb{F}^{(I)}$. ∎

Let the linear spaces $\mathcal{V}$ and $\mathcal{W}$, the family $z \in \mathcal{V}^I$, and the linear mapping $L : \mathcal{V} \to \mathcal{W}$ be given. From Theorem 41E we have

(43.1) $$\mathrm{lc}_{Lz}^{\mathcal{W}} = L\mathrm{lc}_z^{\mathcal{V}}.$$

From (43.1), from the definitions, and from Proposition 43B,(a) it is easy to derive relationships between the properties of z (being linearly independent, spanning $\mathcal{V}$, being a basis of $\mathcal{V}$), analogous properties of Lz with respect to $\mathcal{W}$, and properties of L (being injective, surjective, invertible). For instance, if Lz is linearly independent, so is z; if z is linearly independent and L is injective, then Lz is linearly independent. We shall not make a complete list of these relationships, but we record those that hold when z is a basis of $\mathcal{V}$.

43D. Proposition. *Let the linear mapping $L : \mathcal{V} \to \mathcal{W}$ and the basis $z \in \mathcal{V}^I$ of $\mathcal{V}$ be given. Then the family Lz is linearly independent if and only if L is injective; Lz spans $\mathcal{W}$ if and only if L is surjective; Lz is a basis of $\mathcal{W}$ if and only if L is invertible.*

43E. Proposition. *Let the linear space $\mathcal{V}$ and the set I be given. The following statements are equivalent.*

(i): *There exists a basis $z \in \mathcal{V}^I$ of $\mathcal{V}$.*

(ii): *There exists a basis-set of $\mathcal{V}$ that is equinumerous with I.*

(iii): *$\mathcal{V}$ is linearly isomorphic to $\mathbb{F}^{(I)}$.*

Proof. (i) *implies* (ii). Choose a basis $z \in \mathcal{V}^I$ of $\mathcal{V}$. Then $\mathrm{Rng}z$ is a basis-set of $\mathcal{V}$ and $z|^{\mathrm{Rng}z} : I \to \mathrm{Rng}z$ is a bijection (Proposition 43B,(b)).

(ii) *implies* (i). Choose a basis-set $\mathcal{A}$ of $\mathcal{V}$ that is equinumerous with I. Then we may choose a bijection from I to $\mathcal{A}$, i.e., an injective family $z \in \mathcal{V}^I$ with $\mathrm{Rng}z = \mathcal{A}$. By Proposition 43A,(b), z is a basis of $\mathcal{V}$.

(i) *implies* (iii): Choose a basis $z \in \mathcal{V}^I$ of $\mathcal{V}$. Then $\mathrm{lc}_z^{\mathcal{V}} : \mathbb{F}^{(I)} \to \mathcal{V}$ is a linear isomorphism (Proposition 43B,(a)).

(iii) *implies* (i): Choose a linear isomorphism $L : \mathcal{V} \to \mathbb{F}^{(I)}$. Then the family $L^{-1}\delta^I \in \mathcal{V}^I$ is a basis of $\mathcal{V}$ (Example 43C,(b) and Proposition 43D). ∎

Probably the most important property of bases is described by the characterization in the following theorem.

43F. Theorem. *Let the linear space $\mathcal{V}$ and the family $z \in \mathcal{V}^I$ be given.*

(a): *z is linearly independent and $\mathrm{Lsp}\,\mathrm{Rng}\,z$ has a supplement in $\mathcal{V}$ if and only if for every linear space $\mathcal{W}$ and family $y \in \mathcal{W}^I$ there is at least one solution of*

$$(43.2) \qquad\qquad\qquad ?L \in \mathrm{Lin}(\mathcal{V}, \mathcal{W}), \quad Lz = y.$$

(b): *z spans $\mathcal{V}$ if and only if for every linear space $\mathcal{W}$ and family $y \in \mathcal{W}^I$ there is at most one solution of (43.2).*

(c): *z is a basis of $\mathcal{V}$ if and only if for every linear space $\mathcal{W}$ and family $y \in \mathcal{W}^I$ there is exactly one solution of (43.2).*

Proof. *Proof of* (a). By Theorem 41C and Corollary 21I,(a), z is linearly independent and $\mathrm{Lsp}\,\mathrm{Rng}\,z$ has a supplement in $\mathcal{V}$ if and only if $\mathrm{lc}_z^{\mathcal{V}}$ is linearly left-invertible.

Assume first that $\mathrm{lc}_z^{\mathcal{V}}$ is linearly left-invertible, and choose a linear left-inverse $K \in \mathrm{Lin}(\mathcal{V}, \mathbb{F}^{(I)})$ of $\mathrm{lc}_z^{\mathcal{V}}$. For a given linear space $\mathcal{W}$ and family $y \in \mathcal{W}^I$ we set $L := \mathrm{lc}_y^{\mathcal{W}} K \in \mathrm{Lin}(\mathcal{V}, \mathcal{W})$ and find, using (41.2),

$$Lz = \mathrm{lc}_y^W K \mathrm{lc}_z^V \delta^I = \mathrm{lc}_y^W \delta^I = y,$$

as desired.

To prove the converse, we apply the condition in the assertion to $W := \mathbb{F}^{(I)}$ and $y := \delta^I$. We may then choose $K \in \mathrm{Lin}(V, \mathbb{F}^{(I)})$ such that $Kz = \delta^I$, and therefore $K \mathrm{lc}_z^V \delta^I = \delta^I = 1_{\mathbb{F}^{(I)}}$ of I. By Proposition 41A (cf. Example 43C,(b)), this implies $K \mathrm{lc}_z^V = 1_{\mathbb{F}^{(I)}}$. Thus lc_z^V is linearly left-invertible.

Proof of (b). Assume first that z spans V. For given linear space W, family $y \in W^I$, and linear mappings $L, M \in \mathrm{Lin}(V, W)$ we have the chain of implications

$$Lz = y = Mz \;\Rightarrow\; (M - L)z = 0 \;\Rightarrow\; \mathrm{Rng}\,z \subset \mathrm{Null}(M - L) \;\Rightarrow$$
$$\Rightarrow\; V = \mathrm{Lsp}\,\mathrm{Rng}\,z \subset \mathrm{Null}(M - L) \;\Rightarrow\; M - L = 0 \;\Rightarrow\; L = M;$$

thus (43.2) has at most one solution for given W and $y \in W^I$.

To prove the converse, we apply the condition in the assertion to each linear space W and to the family $y := 0 \in W^I$. Then we have, for every $L \in \mathrm{Lin}(V, W)$, the chain of implications

$$L \mathrm{lc}_z^V = 0 \quad\Rightarrow\quad Lz = L \mathrm{lc}_z^V \delta^I = 0 \quad\Rightarrow\quad Lz = 0z \quad\Rightarrow\quad L = 0.$$

This shows that lc_z^V is a linear epimorphism, and is therefore surjective (Proposition 22B.R). Consequently, z spans V (Proposition 43B,(a)).

Proof of (c). This follows from (a) and (b) and the following remark: when z spans V, then $\mathrm{Lsp}\,\mathrm{Rng}\,z = V$ has the supplement $\{0\}$ in V. ∎

43G. Remarks. (a): If z is a basis of V, the only solution of (43.2) is given by the formula $L := \mathrm{lc}_y^W (\mathrm{lc}_z^V)^{-1}$.

(b): There is an interesting alternative proof of Theorem 43F,(c) that uses Proposition 43B,(d) and the defining property of decompositions. ∎

•**43H. Corollary.** *Let the linear space V and the family $z \in V^I$ be given. z is linearly independent if and* •*only if for every linear space W and family $y \in W^I$ there is at least one solution of* (43.2).

Proof. This follows from Theorem 43F,(a) and •Corollary 15G. ∎

Of the special form of Theorem 43F for (self-indexed) subsets of V we record only the part corresponding to (c).

43I. Corollary. *Let the linear space V and the subset A of V be given. A is a basis-set of V if and only if for every linear space W and mapping $f : A \to W$ there is exactly one linear mapping $L : V \to W$ such that $L|_A = f$.*

Our next results demonstrate that basis-sets are both maximal linearly independent sets and minimal spanning sets, and permit us to discuss the *existence* of basis sets.

43J. Theorem. *Let the linear space V and the subsets A, B, C of V be given, and assume that $A \subset B \subset C$. The following statements are equivalent.*

(i): $\mathcal{B}$ is a basis-set of $\mathcal{V}$.

(ii): $\mathcal{A}$ is linearly independent, $\mathcal{C}$ spans $\mathcal{V}$, and $\mathcal{B}$ is a maximal member of $\{\mathcal{S} \in \mathfrak{P}(\mathcal{V}) \mid \mathcal{A} \subset \mathcal{S} \subset \mathcal{C},\ \mathcal{S}$ is linearly independent$\}$, ordered by inclusion.

(iii): $\mathcal{A}$ is linearly independent, $\mathcal{C}$ spans $\mathcal{V}$, and $\mathcal{B}$ is a minimal member of $\{\mathcal{S} \in \mathfrak{P}(\mathcal{V}) \mid \mathcal{A} \subset \mathcal{S} \subset \mathcal{C},\ \mathcal{S}$ spans $\mathcal{V}\}$, ordered by inclusion.

Proof. (i) *implies* (ii) *and* (iii). We assume that $\mathcal{B}$ is a basis-set of $\mathcal{V}$, and define

(43.3)
$$\Gamma := \{\mathcal{S} \in \mathfrak{P}(\mathcal{V}) \mid \mathcal{A} \subset \mathcal{S} \subset \mathcal{C},\ \mathcal{S} \text{ is linearly independent}\}$$
$$\Delta := \{\mathcal{S} \in \mathfrak{P}(\mathcal{V}) \mid \mathcal{A} \subset \mathcal{S} \subset \mathcal{C},\ \mathcal{S} \text{ spans } \mathcal{V}\}.$$

Obviously, $\mathcal{B} \in \Gamma$ and $\mathcal{B} \in \Delta$. By Corollary 42E and Remark 43A, $\mathcal{A}$ is linearly independent and $\mathcal{C}$ spans $\mathcal{V}$.

Let $\mathcal{B}' \in \Gamma$ be given, with $\mathcal{B} \subset \mathcal{B}'$. Then $\mathcal{B}'$ spans $\mathcal{V}$ (Remark 43A) and is linearly independent. By Corollary 42J,$((i) \Rightarrow (iv))$ we find

$$\mathcal{B} \in \{\mathcal{S} \in \mathfrak{P}(\mathcal{B}') \mid \mathcal{A} \subset \mathcal{S},\ \mathrm{Lsp}\mathcal{S} = \mathcal{V} = \mathrm{Lsp}\mathcal{B}'\} = \{\mathcal{B}'\}.$$

Therefore $\mathcal{B}' = \mathcal{B}$. This shows that $\mathcal{B}$ is a maximal member of Γ.

Let $\mathcal{B}'' \in \Delta$ be given with $\mathcal{B}'' \subset \mathcal{B}$. By Corollary 42J,$((i) \Rightarrow (iv))$ we find

$$\mathcal{B}'' \in \{\mathcal{S} \in \mathfrak{P}(\mathcal{B}) \mid \mathcal{A} \subset \mathcal{S},\ \mathrm{Lsp}\mathcal{S} = \mathcal{V} = \mathrm{Lsp}\mathcal{B}\} = \{\mathcal{B}\}.$$

Therefore $\mathcal{B}'' = \mathcal{B}$. This shows that $\mathcal{B}$ is a minimal member of Δ.

(ii) *implies* (i). We assume that (ii) holds. Then $\mathcal{B}$ is linearly independent. Let $v \in \mathcal{C}$ be given. If we had $v \notin \mathrm{Lsp}\mathcal{B}$, we should have $\mathcal{A} \subset \mathcal{B} \subset \mathcal{B} \cup \{v\} \subset \mathcal{C}$, and by Corollary 42J,$((ii) \Rightarrow (i))$ $\mathcal{B} \cup \{v\}$ would be linearly independent, contradicting the assumed maximality of $\mathcal{B}$ in Γ (defined by (43.3)). Therefore $v \in \mathrm{Lsp}\mathcal{B}$. Since $v \in \mathcal{C}$ was arbitrary, we find $\mathcal{C} \subset \mathrm{Lsp}\mathcal{B}$, and hence $\mathcal{V} = \mathrm{Lsp}\mathcal{C} \subset \mathrm{Lsp}\mathcal{B} \subset \mathcal{V}$, so that $\mathcal{B}$ spans $\mathcal{V}$. Since $\mathcal{B}$ was also linearly independent, $\mathcal{B}$ is a basis-set of $\mathcal{V}$.

(iii) *implies* (i). We assume that (iii) holds. Then $\mathcal{A}$ is linearly independent and, since $\mathcal{B}$ is a minimal member of Δ (defined by (43.3)), $\mathcal{B}$ is the only member of $\{\mathcal{S} \in \mathfrak{P}(\mathcal{B}) \mid \mathcal{A} \subset \mathcal{S},\ \mathrm{Lsp}\mathcal{S} = \mathcal{V} = \mathrm{Lsp}\mathcal{B}\}$. By Corollary 42J,$((iv) \Rightarrow (i))$, $\mathcal{B}$ is linearly independent. Since $\mathcal{B} \in \Delta$, $\mathcal{B}$ also spans $\mathcal{V}$. Therefore $\mathcal{B}$ is a basis-set of $\mathcal{V}$. ∎

43K. Corollary. *Let the linear space $\mathcal{V}$ and the subset $\mathcal{B}$ of $\mathcal{V}$ be given. The following statements are equivalent.*

(i): $\mathcal{B}$ is a basis-set of $\mathcal{V}$.

(ii): $\mathcal{B}$ is a maximal member of the collection of all linearly independent subsets of $\mathcal{V}$, ordered by inclusion.

(iii): $\mathcal{B}$ is a minimal member of the collection of all subsets of $\mathcal{V}$ that span $\mathcal{V}$, ordered by inclusion.

Proof. We apply Theorem 43J with $\mathcal{A} := \emptyset$ and $\mathcal{C} := \mathcal{V}$. ∎

•**43L. Theorem.** *Let the linear space* V *and the subsets* A *and* C *of* V *be given, and assume that* $A \subset C$. *There exists a basis-set* B *of* V *with* $A \subset B \subset C$ •*if and only if* A *is linearly independent and* C *spans* V.

Proof. The "only if" part follows at once form Theorem 43J,((i) $\Rightarrow$ (ii)). To prove the • "if" part, we recall that the collection $\Lambda := \{S \in \mathfrak{P}(V) \mid S \text{ is linearly independent}\}$ is of finitary character (Corollary 42E,(b)). We assume that A is linearly independent, i.e., $A \in \Lambda$, and that C spans V. Since $A \subset C$, it follows from •Tukey's Lemma ((VI) in *Basic Language*, Section 173) and *Basic Language*, Proposition 173E that we may choose a maximal member B of $\{S \in \Lambda \mid A \subset S \subset C\} =$ $= \{S \in \mathfrak{P}(V) \mid A \subset S \subset C, S \text{ is linearly independent}\}$. Then $A \subset B \subset C$, and by Theorem 43J,((ii) $\Rightarrow$ (i)) B is a basis-set of V. ∎

•**43M. Corollary.** *Every linear space* V *includes a basis-set of* V. *Every linear space* V *is isomorphic to* $\mathbb{F}^{(I)}$ *for some set* I.

Proof. Apply •Theorem 43L to $A := \emptyset$ and $C := V$. The last assertion follows by Proposition 43E. ∎

▼ •Corollary 43M answers the question regarding the existence of basis-sets by showing that every linear space has at least one. By contrast, only the most degenerate linear spaces have exactly one basis-set, as we now show.

43N. Proposition. *A linear space* V *has exactly one basis set if and only if either* V *is a zero-space or else the set* $\mathbb{F}$ *is the doubleton* $\{0, 1\}$ *and* V *is linearly isomorphic to the linear space* $\mathbb{F}$.

Proof. Since every linear space V with a basis-set is linearly isomorphic to $\mathbb{F}^{(I)}$ for some set I (Proposition 43E), we may restrict our attention to the spaces $\mathbb{F}^{(I)}$, each with its Kronecker basis δ^I (Example 43C,(b)).

Consider first a set I that is neither empty nor a singleton, and choose distinct $j, k \in I$. Define the linear mappings $K, L \in \text{Lin}(\mathbb{F}^{(I)}, \mathbb{F}^{(I)})$ by requiring (as we may, by Proposition 41A (cf. Theorem 43F,(c)),

$$K\delta_i^I := \begin{cases} \delta_j^I + \delta_k^I \\ \delta_i^I \end{cases} \qquad L\delta_i^I := \begin{cases} \delta_j^I - \delta_k^I & \text{if } i = j \\ \delta_i^I & \text{if } i \in I \setminus \{j\} \end{cases}$$

Then $KL\delta^I = \delta^I = LK\delta^I$, and, again by Proposition 41A, we infer that $KL = 1_{\mathbb{F}^{(I)}} = LK$. It follows from Propositions 43D and 43B,(b) that $\text{Rng}(K\delta^I)$ is a basis-set of $\mathbb{F}^{(I)}$; but it is different from the basis-set $\text{Rng}\delta^I$, since the support of $K\delta_j^I = \delta_j^I + \delta_k^I$ is not a singleton. Therefore $\mathbb{F}^{(I)}$ has more bases than one.

The set I is empty if and only if $\mathbb{F}^{(I)}$ is a zero-space; and then $\emptyset$ is the only basis-set of $\mathbb{F}^{(I)}$.

Assume, finally, that I is a singleton, and define $i :\in I$. Then $\delta_i^I \otimes : \mathbb{F} \to \mathbb{F}^{(I)}$ is a linear isomorphism. But the basis-sets of the linear space $\mathbb{F}$ are precisely the singletons $\{t\}$ with $t \in \mathbb{F}^\times$ (Example 43C,(a)). Therefore $\mathbb{F}$, and also $\mathbb{F}^{(I)}$, has exactly one basis-set if and only if $\mathbb{F}^\times = \{1\}$, i.e., $\mathbb{F} = \{0, 1\}$. ∎

43O. Remark. Proposition 43N is one of the few exceptions to the rule that the validity of statements in Linear Algebra proper does not depend on the specific nature of the field $\mathbb{F}$ (cf. Proposition 15C,(b)). Had we required, as is often done, that $1 + 1 \neq 0$ in $\mathbb{F}$, the proof of Proposition 43N could have been reduced to the observation that if $\mathcal{A}$ is a non-empty basis-set of $\mathcal{V}$, then $-\mathcal{A}$ is a different basis-set of $\mathcal{V}$. ∎

We next consider unions of subsets of a linear space.

43P. LEMMA. *Let the linear space $\mathcal{V}$ and the subsets $\mathcal{A}$ and $\mathcal{A}'$ of $\mathcal{V}$ be given. If $\mathcal{A} \cap \mathcal{A}' = \emptyset$ and $\mathcal{A} \cup \mathcal{A}'$ is linearly independent, then $\mathrm{Lsp}\mathcal{A} \cap \mathrm{Lsp}\mathcal{A}' = \{0\}$.*

Proof. Let $u \in \mathrm{Lsp}\mathcal{A} \cap \mathrm{ALsp}\mathcal{A}'$ be given. By Corollary 41D we may choose $a \in \mathbb{F}^{(\mathcal{A})}$ and $a' \in \mathbb{F}^{(\mathcal{A}')}$ such that $\mathrm{lc}_{\mathcal{A}}a = u = \mathrm{lc}_{\mathcal{A}'}a'$. Since $\mathcal{A} \cap \mathcal{A}' = \emptyset$, we may define $b \in \mathbb{F}^{(\mathcal{A} \cup \mathcal{A}')}$ by requiring $b|_{\mathcal{A}} := a$ and $b|_{\mathcal{A}'} := -a'$. Then

$$\mathrm{lc}_{\mathcal{A}\cup\mathcal{A}'}b = \mathrm{lc}_{\mathcal{A}}(b|_{\mathcal{A}}) + \mathrm{lc}_{\mathcal{A}'}(b|_{\mathcal{A}'}) = \mathrm{lc}_{\mathcal{A}}a - \mathrm{lc}_{\mathcal{A}'}a' = u - u = 0.$$

Since $\mathcal{A} \cup \mathcal{A}'$ is linearly independent, this requires $b = 0$, and hence $u = \mathrm{lc}_{\mathcal{A}}(b|_{\mathcal{A}}) = 0$. since $u \in \mathrm{Lsp}\mathcal{A} \cap \mathrm{Lsp}\mathcal{A}'$ was arbitrary, we find $\mathrm{Lsp}\mathcal{A} \cap \mathrm{Lsp}\mathcal{A}' = \{0\}$. ∎

43Q. THEOREM. *Let the linear space $\mathcal{V}$ and the family $(\mathcal{A}_i \mid i \in I)$ of subsets of $\mathcal{V}$ be given. The following statements are equivalent.*

(i): *The family $(\mathcal{A}_i \mid i \in I)$ is disjoint and $\bigcup_{i \in I}\mathcal{A}_i$ is linearly independent [is a basis-set of $\mathcal{V}$].*

(ii): *The family $(\mathrm{Lsp}\mathcal{A}_i \mid i \in I)$ is disjunct [is a decomposition of $\mathcal{V}$] and $\mathcal{A}_i$ is linearly independent (equivalently, a basis-set of $\mathrm{Lsp}\mathcal{A}_i$) for each $i \in I$.*

Proof. The equivalence in parentheses follows from Proposition 43B,(c). By Corollary 13S, we have $\mathrm{Lsp}\bigcup_{i \in I}\mathcal{A}_i = \sum_{i \in I}\mathrm{Lsp}\mathcal{A}_i$. It follows that we may reduce the assertion to the assertion with the phrases in square brackets by replacing $\mathcal{V}$ by this subspace. When this is done, we have the additional assumption

$$(43.4) \qquad\qquad \mathcal{V} = \mathrm{Lsp}\bigcup_{i \in I}\mathcal{A}_i = \sum_{i \in I}\mathrm{Lsp}\mathcal{A}_i.$$

(i) *implies* (ii). Let $j \in I$ be given. Since $\bigcup_{i \in I}\mathcal{A}_i$ is linearly independent, so is the subset $\mathcal{A}_j$ (Corollary 42E,(a)). Since the family $(\mathcal{A}_i \mid i \in I)$ is disjoint, we have $\mathcal{A}_j \cap \bigcup_{i \in I\setminus\{j\}}\mathcal{A}_i = \emptyset$. By Lemma 43P and Corollary 13S we have

$$\mathrm{Lsp}\mathcal{A}_j \cap \sum_{i \in I\setminus\{j\}}\mathrm{Lsp}\mathcal{A}_i = \mathrm{Lsp}\mathcal{A}_j \cap \mathrm{Lsp}\bigcup_{i \in I\setminus\{j\}}\mathcal{A}_i = \{0\}.$$

This holds for all $j \in I$. It follows from this and (43.4) that $(\mathrm{Lsp}\mathcal{A}_i \mid i \in I)$ is a decomposition of $\mathcal{V}$ (Theorem 33A).

(ii) *implies* (i). Since $\mathcal{A}_i$ is linearly independent, we have $0 \notin \mathcal{A}_i$ for each $i \in I$. Since the family $(\mathrm{Lsp}\mathcal{A}_i \mid i \in I)$ is a decomposition of $\mathcal{V}$, hence is disjunct, it follows that the family $(\mathcal{A}_i \mid i \in I)$ is disjoint.

We shall use the family of projections $(P_i \mid i \in I) \in \bigtimes_{i \in I} \mathrm{Lin}(\mathcal{V}, \mathrm{Lsp}\mathcal{A}_i)$ defined (cf. (33.4)) by requiring

$$(43.5) \qquad P_j|_{\mathrm{Lsp}\mathcal{A}_i} := \begin{cases} 1_{\mathrm{Lsp}\mathcal{A}_j} & \text{if } i = j \\ \\ 0 & \text{if } i \neq j \end{cases} \qquad \text{for all } i, j \in I.$$

Set $\mathcal{B} := \bigcup_{i \in I} \mathcal{A}_i$. Let $b \in \mathrm{Null}\ \mathrm{lc}^{\mathcal{V}}_{\mathcal{B}}$ and $j \in I$ be given. Then, by (43.5),

$$0 = P_j 0 = P_j \mathrm{lc}_{\mathcal{B}} b = P_j \sum_{v \in \mathcal{B}} b_v v = \sum_{v \in \mathcal{B}} b_v P_j v = \sum_{v \in \mathcal{A}_j} b_v v = \mathrm{lc}_{\mathcal{A}_j}(b|_{\mathcal{A}_j}).$$

Since $\mathcal{A}_j$ is linearly independent, it follows that $b|_{\mathcal{A}_j} = 0$. Since $j \in I$ was arbitrary, we infer that $b = 0$. Since $b \in \mathrm{Null}\ \mathrm{lc}^{\mathcal{V}}_{\mathcal{B}}$ was arbitrary, we conclude that $\mathrm{Null}\ \mathrm{lc}^{\mathcal{V}}_{\mathcal{B}} = \{0\}$, and therefore $\mathcal{B}$ is linearly independent (Remark 42A,(b)). By (43.4), $\mathcal{B} = \bigcup_{i \in I} \mathcal{A}_i$ spans $\mathcal{V}$, and hence is a basis-set of $\mathcal{V}$. ∎

43R. Corollary. *Let the linear space $\mathcal{V}$ and the subsets $\mathcal{A}$ and $\mathcal{A}'$ of $\mathcal{V}$ be given. Then $\mathcal{A} \cap \mathcal{A}' = \varnothing$ and $\mathcal{A} \cup \mathcal{A}'$ is linearly independent [a basis-set of $\mathcal{V}$] if and only if $\mathcal{A}$ and $\mathcal{A}'$ are linearly independent and $\mathrm{Lsp}\mathcal{A}$ and $\mathrm{Lsp}\mathcal{A}'$ are disjunct [supplementary in $\mathcal{V}$].*

43S. Corollary. *Let the family of linear spaces $(\mathcal{V}_i \mid i \in I)$ be given. Let $\mathcal{S}$ be the coproduct-space, and $(s_i \mid i \in I)$ the family of insertions, of a linear coproduct of $(\mathcal{V}_i \mid i \in I)$. For each $i \in I$, let a basis $b_i \in \mathcal{V}_i{}^{J_i}$ be given. Set $U := \bigcup_{i \in I} J_i$. Then the family $c := (s_i(b_i)_j \mid (i, j) \in U) \in \mathcal{S}^U$ is a basis of $\mathcal{S}$.*

Proof. We set $\mathcal{A}_i := \mathrm{Rng}(s_i b_i)$ for each $i \in I$. Since b_i is a basis, we have, by Corollary 13Q,

$$\mathrm{Lsp}\mathcal{A}_i = \mathrm{Lsp}\ \mathrm{Rng}(s_i b_i) = \mathrm{Lsp}(s_i)_{>}(\mathrm{Rng}b_i) = (s_i)_{>}(\mathrm{Lsp}\ \mathrm{Rng}\ b_i) = (s_i)_{>}(\mathcal{V}_i) = \mathrm{Rng}s_i$$
$$\text{for all } i \in I.$$

It follows from Corollary 33H that $(\mathrm{Lsp}\mathcal{A}_i \mid i \in I)$ is a decomposition of $\mathcal{S}$.

Since $s_i|^{\mathrm{Rng}}$ is invertible, it follows that $s_i b_i$ is a basis of $\mathrm{Rng}s_i = \mathrm{Lsp}\mathcal{A}_i$; hence $\mathcal{A}_i = \mathrm{Rng}(s_i b_i)$ is a basis-set of $\mathrm{Lsp}\mathcal{A}_i$ for each $i \in I$. By Theorem 43Q, the family $(\mathcal{A}_i \mid i \in I)$ is disjoint and $\bigcup_{i \in I} \mathcal{A}_i = \mathrm{Rng}c$ is a basis-set of $\mathcal{S}$. Since the families $s_i b_i$ are injective and have disjoint ranges, we conclude that c is injective. It follows from Proposition 43B,(b) that c is a basis of $\mathcal{S}$. ∎

Chapter 5

FINITE DIMENSION

51. Matrices

This section deals with linear mappings from $\mathbb{F}^{(I)}$ to $\mathbb{F}^{(J)}$ for given sets I and J. The $J \times I$-matrices with entries from $\mathbb{F}$ provide a useful bookkeeping system for these mappings, especially when J is finite. To each $L \in \operatorname{Lin}(\mathbb{F}^{(I)}, \mathbb{F}^{(J)})$ one can associate the $J \times I$-matrix $((L\delta_i^I)_j \mid (j,i) \in J \times I)$. It is not hard to see that the mapping from $\operatorname{Lin}(\mathbb{F}^{(I)}, \mathbb{F}^{(J)})$ to $\mathbb{F}^{J \times I}$ thus defined is linear and injective, and that its range consists of all members of $\mathbb{F}^{J \times I}$ whose columns have finite support. When J is finite, this mapping is therefore actually a linear isomorphism. Our first result records this state of affairs.

51A. PROPOSITION. *Let the sets I and J be given, and assume that J is finite. The mappings*

$$(51.1) \qquad (L \mapsto ((L\delta_i^I)_j \mid (j,i) \in J \times I)) : \operatorname{Lin}(\mathbb{F}^{(I)}, \mathbb{F}^J) \to \mathbb{F}^{J \times I}$$

$$(51.2) \qquad (M \mapsto (a \mapsto (\sum_{i \in I} M_{j,i} a_i \mid j \in J))) : \mathbb{F}^{J \times I} \to \operatorname{Lin}(\mathbb{F}^{(I)}, \mathbb{F}^J)$$

are linear, and each is the inverse of the other.

Proof. Let the mappings defined by (51.1) and (51.2) be denoted by Φ and Ψ, respectively. These mappings are obviously linear. Moreover,

$$((\Psi\Phi L)\delta_i^I)_j = \sum_{i' \in I} (\Phi L)_{j,i'} \delta_{i,i'}^I = (\Phi L)_{j,i} = (L\delta_i^I)_j \quad \text{for all } i \in I, j \in J,$$

$$\text{and } L \in \operatorname{Lin}(\mathbb{F}^{(I)}, \mathbb{F}^J).$$

Thus, for every $L \in \operatorname{Lin}(\mathbb{F}^{(I)}, \mathbb{F}^J)$ we have $(\Psi\Phi L)\delta_i^I = L\delta_i^I$ for all $i \in I$; by Proposition 41A, this implies $\Psi\Phi L = L$. Therefore $\Psi\Phi = 1_{\operatorname{Lin}(\mathbb{F}^{(I)}, \mathbb{F}^J)}$. On the other hand,

73

$$(\Phi\Psi M)_{j,i} = ((\Psi M)\delta_i^I)_j = \sum_{i'\in I} M_{j,i'}\delta_{i,i'}^I = M_{j,i} \quad \text{for all } (j,i) \in J \times I \text{ and}$$

$$M \in \mathbb{F}^{J\times I},$$

and therefore $\Phi\Psi = 1_{\mathbb{F}^{J\times I}}$. ∎

The natural linear isomorphism described in Proposition 51A allows us to *identify* $\text{Lin}(\mathbb{F}^{(I)},\mathbb{F}^J)$ and $\mathbb{F}^{J\times I}$, i.e., to use the same symbol to denote a member of $\text{Lin}(\mathbb{F}^{(I)},\mathbb{F}^J)$ and the corresponding $J\times I$-matrix. We record some formulas expressing this identification, derived from (51.1), (51.2), and (13.4). For every M in $\text{Lin}(\mathbb{F}^{(I)},\mathbb{F}^J)$, or in $\mathbb{F}^{J\times I}$, we have

(51.3) $$M_{j,i} = (M\delta_i^I)_j \quad \text{for all } (j,i) \in J \times I,$$

(51.4) $$M\delta_i^I = \sum_{j\in J} M_{j,i}\delta_j^J \quad \text{for all } i \in I,$$

(51.5) $$(Ma)_j = \sum_{i\in I} M_{j,i}a_i \quad \text{for all } j \in J \text{ and } a \in \mathbb{F}^{(I)}.$$

The next result shows that, under this identification, the composition of linear mappings corresponds to the familiar "row-by-column" multiplication of matrices; it is indeed, this result that provides the reason for considering this multiplication at all.

51B. Proposition. *Let the sets I, J, K be given, and assume that J and K are finite. Let $M \in \text{Lin}(\mathbb{F}^{(I)},\mathbb{F}^J)$ and $N \in \text{Lin}(\mathbb{F}^J,\mathbb{F}^K)$ be given. Then the $K \times I$-matrix NM corresponding to their composite is given by*

$$(NM)_{k,i} = \sum_{j\in J} N_{k,j}M_{j,i} \quad for \ all \ (k,i) \in K \times I.$$

Proof. By successive application of (51.3), (51.5), (51.3) we obtain

$$(NM)_{k,i} = (NM\delta_i^I)_k = \sum_{j\in J} N_{k,j}(M\delta_i^I)_j = \sum_{j\in J} N_{k,j}M_{j,i} \quad for \ all \ (k,i) \in K \times I. \ ∎$$

For given sets I and J, with J finite, and given $M \in \text{Lin}(\mathbb{F}^{(I)},\mathbb{F}^J)$ and $b \in \mathbb{F}^J$, we examine the equation

$$?a \in \mathbb{F}^{(I)}, \quad Ma = b,$$

i.e.,

$$?(a_i \mid i \in I) \in \mathbb{F}^{(I)}, \quad (\forall j \in J, \quad \sum_{i \in I} M_{j,i} a_i = b_j);$$

in this form, the problem is usually called a *system of linear equations* (more precisely, of #J *linear equations in* #I *unknowns*, if I is also finite). The following theorem describes a procedure that enables one to "reduce" such a system to another of analogous form with one equation and one "unknown" fewer. This procedure typically occurs as a step in a recursive process, or algorithm, known as *linear elimination*, for solving a system of linear equations.

51C. Theorem. *Let the sets I and J be given, and assume that J is finite. Let $M \in \mathbb{F}^{J \times I}$ and $b \in \mathbb{F}^J$ be given. Let $(j_0, i_0) \in J \times I$ be such that $M_{j_0, i_0} \neq 0$. Set $I' := I \setminus \{i_0\}$ and $J' := J \setminus \{j_0\}$, and define $M' \in \mathbb{F}^{J' \times I'}$ and $b' \in \mathbb{F}^{J'}$ by the rules*

$$(51.6) \qquad M'_{j,i} := M_{j,i} - M_{j,i_0} M_{j_0,i} / M_{j_0,i_0} \quad for\ all\ (j, i) \in J' \times I'$$

$$(51.7) \qquad b'_j := b_j - (M_{j,i_0} / M_{j_0,i_0}) b_{j_0} \quad for\ all\ j \in J'$$

Then $a \in \mathbb{F}^{(I)}$ is a solution of

$$?a \in \mathbb{F}^{(I)}, \quad Ma = b$$

if and only if $a|_{I'}$ is a solution of

$$?a' \in \mathbb{F}^{(I')}, \quad M'a' = b'$$

and

$$a_{i_0} = (1/M_{j_0,i_0})(b_{j_0} - \sum_{i \in I'} M_{j_0,i} a_i).$$

Proof. For every $a \in \mathbb{F}^{(I)}$ we have, by (51.6) and (51.5),

$$(Ma)_{j_0} = \sum_{i \in I} M_{j_0,i} a_i = M_{j_0,i_0}\left(a_{i_0} + (1/M_{j_0,i_0})\sum_{i \in I'} M_{j_0,i} a_i\right)$$

$$(Ma)_j = \sum_{i \in I} M_{j,i} a_i = \sum_{i \in I'} M_{j,i} a_i + M_{j,i_0} a_{i_0} =$$

$$= \sum_{i \in I'} M'_{j,i} a_i + (M_{j,i_0}/M_{j_0,i_0})\sum_{i \in I'} M_{j_0,i} a_i + M_{j,i_0} a_{i_0} =$$

$$= (M'(a|_{I'}))_j + M_{j,i_0}\left(a_{i_0} + (1/M_{j_0,i_0})\sum_{i \in I'} M_{j_0,i} a_i\right) \quad for\ all\ j \in J'.$$

The assertion follows from these formulas by direct verification, using (51.7). ∎

51D. Corollary. *Let the sets I and J be given, and assume that J is finite. Let $M \in \mathbb{F}^{J \times I}$ be given, and let $(j_0, i_0) \in J \times I$ be such that $M_{j_0,i_0} \neq 0$. Set $I' := I\setminus\{i_0\}$ and $J' := J\setminus\{j_0\}$, and define $M' \in \mathbb{F}^{J' \times I'}$ by (51.6). Then $M' \in \mathrm{Lin}(\mathbb{F}^{(I')}, \mathbb{F}^{J'})$ is injective if and only if $M \in \mathrm{Lin}(\mathbb{F}^{(I)}, \mathbb{F}^{J})$ is injective, and M' is surjective if and only if M is surjective.*

Proof. Assume first that M is injective. Let $a' \in \mathrm{Null}M'$ be given, so that $M'a' = 0$. Define $a \in \mathbb{F}^{(I)}$ by requiring $a|_{I'} := a'$ and $a_{i_0} := -(1/M_{j_0,i_0})\sum_{i\in I'} M_{j_0,i} a'_i$. By Theorem 51C with $b := 0$, we have $Ma = 0$, so that $a = 0$ since M is injective. Therefore $a' = a|_{I'} = 0$. Since $a' \in \mathrm{Null}M'$ was arbitrary, we have $\mathrm{Null}M' = \{0\}$, so that M' is injective. The reverse implication is an obvious consequence of Theorem 51C with $b := 0$.

Assume that M is surjective. Let $b' \in \mathbb{F}^{J'}$ be given. Define $b \in \mathbb{F}^J$ by requiring $b|_{J'} := b'$ and $b_{j_0} := 0$. We may then choose $a \in \mathbb{F}^{(I)}$ such that $Ma = b$. By Theorem 51C we have $M'(a|_{I'}) = b'$. Since $b' \in \mathbb{F}^{J'}$ was arbitrary, M' is surjective. The reverse implication is an obvious consequence of Theorem 51C. ∎

51E. Proposition. *Let the sets I_0 and J_0 be given, and assume that J_0 is finite. If there exists an injective linear mapping from $\mathbb{F}^{(I_0)}$ to $\mathbb{F}^{J_0}$, then I_0 is finite and $\#I_0 \leq \#J_0$.*

Proof. We shall prove by general induction that

$$P(J) :\Leftrightarrow \text{(For every } I \in \mathfrak{P}(I_0), \text{ if there is an injective } M \in \mathrm{Lin}(\mathbb{F}^{(I)}, \mathbb{F}^{J}),$$
$$\text{then } I \text{ is finite and } \#I \leq \#J)$$

holds for all (finite) subsets J of J_0. Once this is accomplished, the assertion follows from $P(J_0)$ with $I := I_0$.

Let the $J \in \mathfrak{P}(J_0)$ be given, and assume that $P(K)$ holds for every proper subset K of J. Let $I \in \mathfrak{P}(I_0)$ be given, and assume that there exists an injective $M \in \mathrm{Lin}(\mathbb{F}^{(I)}, \mathbb{F}^{J})$; choose such an M. If $M = 0$, then $\mathbb{F}^{(I)} = \mathrm{Null}M = \{0\}$, and hence $I = \emptyset$, so that indeed I is finite $\#I = 0 \leq \#J$ in this case. We may therefore suppose that $M \neq 0$, and choose $(j_0, i_0) \in J \times I$ such that $M_{j_0,i_0} \neq 0$. We set $I' := I\setminus\{i_0\}$ and $J' := J\setminus\{j_0\}$. It follows from Corollary 51D that there is an injective $M' \in \mathrm{Lin}(\mathbb{F}^{(I')}, \mathbb{F}^{J'})$. By the induction hypothesis $P(J')$ we conclude that $I' = I\setminus\{i_0\}$ is finite and that $\#I' \leq \#J'$. Therefore I is finite and

$$\#I = \#(I\setminus\{i_0\}) + 1 = \#I' + 1 \leq \#J' + 1 = \#(J\setminus\{j_0\}) + 1 = \#J.$$

This completes the induction step. ∎

52. Finite-dimensional spaces

A linear space $\mathcal{V}$ is said to be **finite-dimensional** if $\mathcal{V}$ is the linear span of some finite subset of $\mathcal{V}$. A linear space is said to be **infinite-dimensional** if it is not finite-dimensional. If $\mathcal{V}$ is a finite-dimensional linear space, the natural number

$$(52.1) \qquad \dim\mathcal{V} := \min\{\#\mathcal{S} \mid \mathcal{S} \in \mathfrak{F}(\mathcal{V}),\ \mathrm{Lsp}\mathcal{S} = \mathcal{V}\}$$

is called the **dimension of** $\mathcal{V}$; for given $n \in \mathbb{N}$, a linear space $\mathcal{V}$ is said to be **n-dimensional** if $\mathcal{V}$ is finite-dimensional and $\dim\mathcal{V} = n$.

(A linear space spanned by a finite subset might more aptly be termed *finitely spanned*. The terms "finite-dimensional" and "infinite-dimensional" are, however, in practically exclusive use, and suggest that "dimension" is defined for *every* linear space in such a way that this "dimension" is "finite" if and only if the space is spanned by a finite subset. Such a definition can be arranged, say by setting $\dim\mathcal{V} := \infty$ for every infinite-dimensional linear space $\mathcal{V}$ (and cf. Remark 54F); but it has little to do with the central and immediate import of the notion of finite-dimensionality.)

52A. Theorem. *Let the finite-dimensional linear space $\mathcal{V}$ be given.*

(a): *If $\mathcal{A}$ is a linearly independent subset of $\mathcal{V}$, then $\mathcal{A}$ is finite and $\#\mathcal{A} \leq \dim\mathcal{V}$.*

(b): *If $\mathcal{A}$ is a finite subset of $\mathcal{V}$ that spans $\mathcal{V}$, then $\#\mathcal{A} \geq \dim\mathcal{V}$.*

(c): *Let the subset $\mathcal{A}$ of $\mathcal{V}$ be given. The following statements are equivalent.*

    (i): *$\mathcal{A}$ is a basis-set of $\mathcal{V}$; i.e., $\mathcal{A}$ is linearly independent and spans $\mathcal{V}$.*

    (ii): *$\mathcal{A}$ is finite and linearly independent, and $\#\mathcal{A} = \dim\mathcal{V}$.*

    (iii): *$\mathcal{A}$ is finite and spans $\mathcal{V}$, and $\#\mathcal{A} = \dim\mathcal{V}$.*

(d): *$\mathcal{V}$ includes a basis-set of $\mathcal{V}$.*

Proof. 1. The assertion (b) follows from (52.1), the definition of $\dim\mathcal{V}$.

Let $\mathcal{A}$ be a finite subset of $\mathcal{V}$ that spans $\mathcal{V}$ and satisfies $\#\mathcal{A} = \dim\mathcal{V}$. By (52.1), $\mathcal{A}$ is a minimal member of the collection of all subsets of $\mathcal{V}$ that span $\mathcal{V}$ ordered by inclusion. By Corollary 43K, $\mathcal{A}$ is a basis-set of $\mathcal{V}$. This establishes the implication (iii) $\Rightarrow$ (i) in (c).

2. By (52.1) we may choose a finite subset $\mathcal{B}$ of $\mathcal{V}$ such that $\mathcal{B}$ spans $\mathcal{V}$ and $\#\mathcal{B} = \dim\mathcal{V}$. By the implication that we have just proved, $\mathcal{B}$ is a basis-set of $\mathcal{V}$; this establishes (d).

Let $\mathcal{A}$ be a linearly independent subset of $\mathcal{V}$. The linear mapping $(\mathrm{lc}_{\mathcal{B}}^{\mathcal{V}})^{-1}\mathrm{lc}_{\mathcal{A}}^{\mathcal{V}} : \mathbb{F}^{(\mathcal{A})} \to \mathbb{F}^{\mathcal{B}}$ is injective. By Proposition 51E, we conclude that $\mathcal{A}$ is finite and that $\#\mathcal{A} \leq \#\mathcal{B} = \dim\mathcal{V}$. This establishes (a).

Suppose that $\mathcal{A}$ is linearly independent (and hence finite), with $\#\mathcal{A} = \dim\mathcal{V}$. Then (a) shows that $\mathcal{A}$ is a maximal member of the collection of all linearly independent subsets of $\mathcal{V}$, ordered by inclusion. By Corollary 43K, $\mathcal{A}$ is a basis-set of $\mathcal{V}$. This establishes the implication (ii) $\Rightarrow$ (i) in (c).

3. Let the basis-set $\mathcal{A}$ of $\mathcal{V}$ be given. Since $\mathcal{A}$ is linearly independent, (a) shows that $\mathcal{A}$ is finite and that $\#\mathcal{A} \leq \dim\mathcal{V}$. Since $\mathcal{A}$ is finite and spans $\mathcal{V}$, (b) shows that

$\#\mathcal{A} \geq \dim\mathcal{V}$. Therefore $\#\mathcal{A} = \dim\mathcal{V}$. This establishes the implications (i) $\Rightarrow$ (ii) and (i) $\Rightarrow$ (iii) in (c), and completes the proof. ∎

52B. COROLLARY. *Let the finite-dimensional linear space $\mathcal{V}$ be given.*

(a): *Every linearly independent subset of $\mathcal{V}$ is finite, and*

$$(52.2) \qquad \dim\mathcal{V} = \max\{\#\mathcal{S} \mid \mathcal{S} \in \mathfrak{P}(\mathcal{V}), \mathcal{S} \text{ is linearly independent}\}.$$

(b): *$\mathcal{V}$ includes a basis-set of $\mathcal{V}$; every basis-set of $\mathcal{V}$ is finite and its cardinal is $\dim\mathcal{V}$.*

(c): *If $\mathcal{A}$ is a linearly independent subset of $\mathcal{V}$, then $\mathrm{Lsp}\mathcal{A}$ is finite-dimensional, and $\dim \mathrm{Lsp}\mathcal{A} = \#\mathcal{A}$.*

From Theorem 52A we derive an analogous result for families in a finite-dimensional linear space.

52C. COROLLARY. *Let the finite-dimensional linear space $\mathcal{V}$ and the family $z \in \mathcal{V}^I$ be given.*

(a): *If z is linearly independent, then I is finite and $\#I \leq \dim\mathcal{V}$.*

(b): *If z spans $\mathcal{V}$ and I is finite, then $\#I \geq \dim\mathcal{V}$.*

(c): *The following statements are equivalent.*

　　(i): *z is a basis of $\mathcal{V}$; i.e., z is linearly independent and spans $\mathcal{V}$.*

　　(ii): *z is linearly independent, and (I is finite and) $\#I = \dim\mathcal{V}$.*

　　(iii): *z spans $\mathcal{V}$ and I is finite and $\#I = \dim\mathcal{V}$.*

Proof. 1. Assume first that z is linearly independent. Then z is injective and $\mathrm{Rng}z$ is linearly independent (Proposition 42C). The mapping $z|^{\mathrm{Rng}z} : I \to \mathrm{Rng}z$ is bijective. Hence I is finite if and only if $\mathrm{Rng}z$ is finite, and then $\#I = \#\mathrm{Rng}z$. Moreover, z spans $\mathcal{V}$ if and only if $\mathrm{Rng}z$ spans $\mathcal{V}$. We may then apply Theorem 52A to $\mathcal{A} := \mathrm{Rng}z$ and find that (a) and the implications (ii) $\Leftrightarrow$ (i) $\Rightarrow$ (iii) in (c) hold.

2. Assume now that I is finite and that z spans $\mathcal{V}$, so that $\mathrm{Rng}z$ spans $\mathcal{V}$. Now the mapping $z|^{\mathrm{Rng}z} : I \to \mathrm{Rng}z$ is surjective. Therefore $\mathrm{Rng}z$ is finite and $\#I \geq \mathrm{Rng}z$, with equality holding (if and) only if z is injective (*Basic Language*, Proposition 101G). From Theorem 52A,(b),(c) we find that $\#\mathrm{Rng}z \geq \dim\mathcal{V}$, with equality holding (if and) only if $\mathrm{Rng}z$ is a basis-set of $\mathcal{V}$. Then $\#I \geq \#\mathrm{Rng}z \geq \dim\mathcal{V}$, which establishes (b). Finally, $\#I = \dim\mathcal{V}$ implies that z is injective and $\mathrm{Rng}z$ is a basis-set of $\mathcal{V}$, so that z is a basis of $\mathcal{V}$ (Proposition 43B,(b)). This establishes the implication (iii) $\Rightarrow$ (i) in (c), and completes the proof. ∎

52D. EXAMPLES. (a): The finite-dimensional linear spaces with dimension 0 are precisely the zero-spaces.

(b): The linear space $\mathbb{F}$ is finite-dimensional, with $\dim\mathbb{F} = 1$.

(c): Let the set I be given. Applying Corollary 52C to the Kronecker basis δ^I of $\mathbb{F}^{(I)}$, it follows that $\mathbb{F}^{(I)}$ is finite-dimensional if and only if I is finite; and in that case $\dim\mathbb{F}^{(I)} = \#I$. ∎

We combine a part of Corollary 52B with a converse into a characterization of finite-dimensionality. To that end, we define

(52.3) $\text{Dim}(\mathcal{V}) := \{\#\mathcal{S} \mid \mathcal{S} \in \mathfrak{F}(\mathcal{V}), \ \mathcal{S} \text{ is linearly independent}\}$

for every linear space $\mathcal{V}$.

52E. Proposition. *Let the linear space $\mathcal{V}$ be given.*

(a): $\text{Dim}(\mathcal{V}) = \{\text{dim}\,\mathcal{U} \mid \mathcal{U} \in \text{Subsp}(\mathcal{V}), \ \mathcal{U} \text{ is finite-dimensional}\}$.

(b): *If $\mathcal{V}$ is finite-dimensional, $\text{Dim}(\mathcal{V}) = (1+\text{dim}\mathcal{V})^{\sqsubset}$; if $\mathcal{V}$ is infinite-dimensional,* $\text{Dim}(\mathcal{V}) = \mathbb{N}$.

Proof. *Proof of* (a). By Corollary 52B and Proposition 43B,(c), a subspace $\mathcal{U}$ of $\mathcal{V}$ is finite-dimensional if and only if it is the linear span of some finite linearly independent subset $\mathcal{S}$ of $\mathcal{V}$; and in that case $\text{dim}\,\mathcal{U} = \#\mathcal{S}$, since $\mathcal{S}$ is a basis-set of $\mathcal{U}$.

Proof of (b). We note that $\text{Dim}(\mathcal{V})$ is not empty, since it contains $\#\emptyset = 0$. Every subset of a linearly independent subset of $\mathcal{V}$ is also linearly independent (Corollary 43E). It follows that $\text{Dim}(\mathcal{V}) = (1 + \max \text{Dim}(\mathcal{V}))^{\sqsubset}$ if $\text{Dim}(\mathcal{V})$ is bounded; and that $\text{Dim}(\mathcal{V}) = \mathbb{N}$ if $\text{Dim}(\mathcal{V})$ is unbounded, and in particular if there exists an infinite linearly independent subset of $\mathcal{V}$ (*Basic Language*, Proposition 101N).

Assume that $\text{Dim}(\mathcal{V})$ is bounded; then all linearly independent subsets of $\mathcal{V}$ are finite. We choose one, $\mathcal{A}$, with $\#\mathcal{A} = \max \text{Dim}(\mathcal{V})$. Then $\mathcal{A}$ is obviously a maximal member of the collection of all linearly independent subsets of $\mathcal{V}$, and therefore spans $\mathcal{V}$ (Corollary 43K). Hence $\mathcal{V}$ is finite-dimensional.

Assume, conversely, that $\mathcal{V}$ is finite-dimensional. By Corollary 52B,(a), $\text{Dim}(\mathcal{V})$ has a maximum, namely $\text{dim}\mathcal{V}$. ■

52F. Corollary. *Let the finite-dimensional linear space $\mathcal{V}$ and the subspace of $\mathcal{U}$ of $\mathcal{V}$ be given. Then $\mathcal{U}$ is finite-dimensional, and $\text{dim}\,\mathcal{U} \leq \text{dim}\mathcal{V}$. Moreover,* $\text{dim}\,\mathcal{U} = \text{dim}\mathcal{V}$ *(if and) only if $\mathcal{U} = \mathcal{V}$.*

Proof. From (52.3) it follows that $\text{Dim}(\mathcal{U}) \subset \text{Dim}(\mathcal{V})$. From Proposition 52E it then follows that $\mathcal{U}$ is finite-dimensional and that $\text{dim}\,\mathcal{U} \leq \text{dim}\mathcal{V}$. If $\text{dim}\,\mathcal{U} = \text{dim}\mathcal{V}$, choose a basis-set $\mathcal{A}$ of $\mathcal{U}$; then $\#\mathcal{A} = \text{dim}\,\mathcal{U} = \text{dim}\mathcal{V}$ (Corollary 52B,(b)). But $\mathcal{A}$ is a linearly independent subset of $\mathcal{V}$; by Theorem 52A,(c), $\mathcal{A}$ is a basis-set of $\mathcal{V}$, and hence $\mathcal{V} = \text{Lsp}\mathcal{A} = \mathcal{U}$. ■

52G. Corollary. *Let the finite-dimensional linear space $\mathcal{V}$ be given. Every non-empty subcollection of $\text{Subsp}(\mathcal{V})$ ordered by inclusion has both a minimal and a maximal member; i.e., $\text{Subsp}(\mathcal{V})$ is well-founded both by inclusion and by the reverse of inclusion.*

Proof. By Corollary 52F we may define the mapping

$$\mathcal{U} \mapsto \text{dim}\,\mathcal{U} : \text{Subsp}(\mathcal{V}) \to (1 + \text{dim}\mathcal{V})^{\sqsubset}$$

and infer that this mapping is strictly $\subset$ $-$ $\leq$-isotone. Since every non-empty ordered subset of $(1 + \text{dim}\mathcal{V})^{\sqsubset}$ has both a minimum and a maximum, *Basic Language*, Proposition 62C,(d) yields the assertion. ■

52H. Remark. In two proofs in the preceding chapters, the existence of maximal members of ordered subcollections of $\text{Subsp}(\mathcal{V})$ was established by appealing to some equivalent of the •Axiom of Choice. Corollary 52G shows that, if $\mathcal{V}$ is finite-dimensional, such an appeal is not required. The proofs in question were those of •Theorem 15F (existence of a supplement of a subspace), and •Theorem 43L (existence of a basis-set between a prescribed linearly independent subset and a prescribed spanning subset). When dealing with finite-dimensional spaces, therefore, these theorems, as well as the other results whose proofs relied on them, viz., •Corollaries 15G, 21K, 21L, 21O, and 21P, •Propositions 22F.L and 22F.R, and •Theorem 22G; and •Corollaries 43H and 43M, may be cited without the "bullet" •. ∎

Proposition 52E shows that an infinite-dimensional linear space has finite linearly independent subsets with arbitrarily great cardinal. We now show that, in fact, it has an infinite linearly independent subset.

•**52I. Proposition.** *A linear space $\mathcal{V}$ is finite-dimensional •if and only if every linearly independent subset of $\mathcal{V}$ is finite.*

Proof. The "only if/" part follows from Corollary 52B,(a). To prove the •"if" part, we assume that every linearly independent subset of $\mathcal{V}$ is finite. By •Corollary 43M, we may choose a basis-set $\mathcal{A}$ of $\mathcal{V}$. Since $\mathcal{A}$ is linearly independent, $\mathcal{A}$ is finite; since $\mathcal{A}$ spans $\mathcal{V}$, $\mathcal{V}$ is finite-dimensional.

We present an alternative proof of the •"if" part that relies only on the •Axiom of Countable Choice. It proceeds by contraposition and is modelled on one proof of *Basic Language*, •Theorem 121V. In view of Proposition 52E,(b) it suffices to assume that $\mathrm{Dim}(\mathcal{V}) = \mathbb{N}$ and to infer that there exists an infinite linearly independent subset of $\mathcal{V}$.

Assume then that $\mathrm{Dim}(\mathcal{V}) = \mathbb{N}$. By the •Axiom of Countable Choice we may choose a sequence $(\mathcal{A}_n \mid n \in \mathbb{N})$ of finite linearly independent subsets of $\mathcal{V}$ with $\#\mathcal{A}_n = 2^n$ for every $n \in \mathbb{N}$. For each $n \in \mathbb{N}$ we set $\mathcal{W}_n := \mathrm{Lsp} \bigcup_{k \in n^{\complement}} \mathcal{A}_k$ and observe that $\mathcal{W}_n$ is finite-dimensional, with

$$\mathrm{dim} \mathcal{W}_n \leq \# \bigcup_{k \in n^{\complement}} \mathcal{A}_k \leq \sum_{k \in n^{\complement}} \#\mathcal{A}_k \leq \sum_{k \in n^{\complement}} 2^k = 2^n - 1 < 2^n = \#\mathcal{A}_n;$$

by Theorem 52A,(a) we cannot have $\mathcal{A}_n \subset \mathcal{W}_n$, and therefore $\mathcal{A}_n \backslash \mathcal{W}_n \neq \varnothing$.

Again by the •Axiom of Countable Choice we may choose a sequence $z \in \bigtimes_{n \in \mathbb{N}} (\mathcal{A}_n \backslash \mathcal{W}_n)$. For each $n \in \mathbb{N}$ and all $k \in n^{\complement}$ we have $z_k \in \mathcal{A}_k \subset \mathcal{W}_n$, and therefore $\mathrm{Lsp}\{z_k \mid k \in n^{\complement}\} \subset \mathcal{W}_n$. It follows that $z_n \notin \mathrm{Lsp}\{z_k \mid k \in n^{\complement}\}$ for all $n \in \mathbb{N}$. We apply Corollary 42L and Proposition 42C, and find that the sequence z is linearly independent and therefore $\mathrm{Rng} z$ is an infinite linearly independent subset of $\mathcal{V}$. ∎

We next examine the relationship between decompositions and finite dimension.

52J. Theorem. *Let the linear space $\mathcal{V}$ and the decomposition $(\mathcal{U}_i \mid i \in I)$ of $\mathcal{V}$ be given. Then $\mathcal{V}$ is finite-dimensional if and only if $\mathrm{Supp}(\mathcal{U}_i \mid i \in I)$ is finite and $\mathcal{U}_i$ is finite-dimensional for every $i \in I$. If these conditions are satisfied, then*

$$(52.4) \qquad \dim \mathcal{V} = \sum_{i \in I} \dim \mathcal{U}_i.$$

Proof. Proof of the "if" part. We assume that $S := \mathrm{Supp}(\mathcal{U}_i \mid i \in I)$ is finite and that $\mathcal{U}_i$ is finite-dimensional for every $i \in I$. We may choose a (finite) basis-set $\mathcal{A}_i$ of $\mathcal{U}_i$ for all $i \in S$, and hence (with $\mathcal{A}_i := \emptyset$ for all $i \in I \backslash S$) for all $i \in I$. Then $\#\mathcal{A}_i = \dim \mathcal{U}_i$ for all $i \in I$ (Corollary 52B,(b)) and $\bigcup_{i \in I} \mathcal{A}_i = \bigcup_{i \in S} \mathcal{A}_i$ is finite. By Theorem 43Q, the family $(\mathcal{A}_i \mid i \in I)$ is disjoint and the finite set $\bigcup_{i \in I} \mathcal{A}_i$ is a basis-set of $\mathcal{V}$. It follows that $\mathcal{V}$ is finite-dimensional and, by Corollary 52B,(b),

$$\dim \mathcal{V} = \# \bigcup_{i \in I} \mathcal{A}_i = \sum_{i \in I} \#\mathcal{A}_i = \sum_{i \in I} \dim \mathcal{U}_i,$$

so that (52.4) holds.

Proof of the "only if" part. We assume that $\mathcal{V}$ is finite-dimensional. By Corollary 52F, $\mathcal{U}_i$ is finite-dimensional for every $i \in I$. Let J be a finite subset of $\mathrm{Supp}(\mathcal{U}_i \mid i \in I)$. Then the restriction $(\mathcal{U}_i \mid i \in J)$ of the disjunct family $(\mathcal{U}_i \mid i \in I)$ is also disjunct (Remark 33D,(b)). Therefore $(\mathcal{U}_i \mid i \in J)$ is a decomposition of the subspace $\mathcal{W} := \sum_{i \in J} \mathcal{U}_i$ of $\mathcal{V}$. We observe that $\dim \mathcal{U}_i \geq 1$ for all $i \in J$. By Corollary 52F, $\mathcal{W}$ is finite-dimensional, and $\dim \mathcal{W} \leq \dim \mathcal{V}$. Applying the already-established "if" part to this decomposition, we find

$$(52.5) \qquad \#J = \sum_{i \in J} 1 \leq \sum_{i \in J} \dim \mathcal{U}_i = \dim \mathcal{W} \leq \dim \mathcal{V}.$$

Since J was an arbitrary finite subset of $\mathrm{Supp}(\mathcal{U}_i \mid i \in I)$, we conclude from (52.5) that this support is finite (cf. *Basic Language*, Proposition 101N). ∎

52K. PROPOSITION. *Let the finite-dimensional linear space $\mathcal{V}$ and the family $(\mathcal{U}_i \mid i \in I)$ of subspaces of $\mathcal{V}$ be given, and assume that this family has finite support. The following statements are equivalent.*

(i): *$(\mathcal{U}_i \mid i \in I)$ is a decomposition of $\mathcal{V}$; i.e., $(\mathcal{U}_i \mid i \in I)$ is disjunct and $\sum_{i \in I} \mathcal{U}_i = \mathcal{V}$.*

(ii): *$(\mathcal{U}_i \mid i \in I)$ is disjunct and $\sum_{i \in I} \dim \mathcal{U}_i = \dim \mathcal{V}$ of $\mathcal{V}$.*

(iii): *$\sum_{i \in I} \mathcal{U}_i = \mathcal{V}$ and $\sum_{i \in I} \dim \mathcal{U}_i = \dim \mathcal{V}$.*

Proof. The implications (i) $\Rightarrow$ (ii) and (i) $\Rightarrow$ (iii) follow from Theorem 52J.

(ii) *implies* (i). $(\mathcal{U}_i \mid i \in I)$ is a decomposition of the subspace $\mathcal{W} := \sum_{i \in I} \mathcal{U}_i$

of $\mathcal{V}$. By Theorem 52J we have $\dim\mathcal{W} = \sum_{i \in I} \dim\mathcal{U}_i = \dim\mathcal{V}$. By Corollary 52F we find $\mathcal{V} = \mathcal{W} = \sum_{i \in I} \mathcal{U}_i$.

(iii) *implies* (i). We choose a basis-set $\mathcal{A}_j$ of $\mathcal{U}_j$ for every $j \in \mathrm{Supp}(\mathcal{U}_i \mid i \in I)$, and hence (with $\mathcal{A}_j := \varnothing$ for all other j) for all $j \in I$. By Corollary 13S we have

$$\mathrm{Lsp} \bigcup_{i \in I} \mathcal{A}_i = \sum_{i \in I} \mathrm{Lsp}\mathcal{A}_i = \sum_{i \in I} \mathcal{U}_i = \mathcal{V}.$$

Therefore $\bigcup_{i \in I} \mathcal{A}_i$ spans $\mathcal{V}$. By the assumption and Theorem 52A,(b) we have

$$\sum_{i \in I} \dim\mathcal{U}_i = \dim\mathcal{V} \leq \# \bigcup_{i \in I} \mathcal{A}_i \leq \sum_{i \in I} \#\mathcal{A}_i = \sum_{i \in I} \dim\mathcal{U}_i.$$

Therefore equality must hold at each of the inequality signs. The first of these equalities requires that $\bigcup_{i \in I} \mathcal{A}_i$ be a basis-set of $\mathcal{V}$ (Theorem 52A,(c)). The second equality requires that the family $(\mathcal{A}_i \mid i \in I)$ be disjoint (*Basic Language*, Corollary 103E). It then follows from Theorem 43Q that $(\mathcal{U}_i \mid i \in I)$ is a decomposition of $\mathcal{V}$. ∎

52L. Corollary. *Let the finite-dimensional linear space $\mathcal{V}$ and the subspaces $\mathcal{U}$ and $\mathcal{U}'$ of $\mathcal{V}$ be given. The following statements are equivalent.*

(i): $\mathcal{U}$ *and* $\mathcal{U}'$ *are supplementary in* $\mathcal{V}$; *i.e.,* $\mathcal{U} \cap \mathcal{U}' = \{0\}$ *and* $\mathcal{U} + \mathcal{U}' = \mathcal{V}$.

(ii): $\mathcal{U} \cap \mathcal{U}' = \{0\}$ *and* $\dim\mathcal{U} + \dim\mathcal{U}' = \dim\mathcal{V}$.

(iii): $\mathcal{U} + \mathcal{U}' = \mathcal{V}$ *and* $\dim\mathcal{U} + \dim\mathcal{U}' = \dim\mathcal{V}$.

52M. Proposition. *Let the linear space $\mathcal{V}$ and the finite-dimensional subspaces $\mathcal{U}$ and $\mathcal{U}'$ of $\mathcal{V}$ be given. Then $\mathcal{U} + \mathcal{U}'$ and $\mathcal{U} \cap \mathcal{U}'$ are finite-dimensional, and*

$$(52.6) \qquad \dim(\mathcal{U} + \mathcal{U}') + \dim(\mathcal{U} \cap \mathcal{U}') = \dim\mathcal{U} + \dim\mathcal{U}'.$$

Proof. Choose finite spanning subsets $\mathcal{A}$ in $\mathcal{U}$ and $\mathcal{A}'$ in $\mathcal{U}'$. By Corollary 13S, $\mathcal{U} + \mathcal{U}' = \mathrm{Lsp}\mathcal{A} + \mathrm{Lsp}\mathcal{A}' = \mathrm{Lsp}(\mathcal{A} \cup \mathcal{A}')$, so that $\mathcal{U} + \mathcal{U}'$ is finite-dimensional. The subspace $\mathcal{U} \cap \mathcal{U}'$ of $\mathcal{U} + \mathcal{U}'$ is finite-dimensional, by Corollary 52F.

By Corollary 15G (cf. Remark 52H), we may choose a (finite-dimensional) supplement $\mathcal{W}$ of $\mathcal{U} \cap \mathcal{U}'$ in $\mathcal{U}'$. By Proposition 15E, $\mathcal{W}$ is a supplement of $\mathcal{U}$ in $\mathcal{U} + \mathcal{U}'$. We apply Corollary 52L ((i) $\Rightarrow$ (ii)) once with $\mathcal{U} + \mathcal{U}'$, $\mathcal{U}$, $\mathcal{W}$ and again with $\mathcal{U}'$, $\mathcal{U} \cap \mathcal{U}'$, $\mathcal{W}$ instead of $\mathcal{V}$, $\mathcal{U}$, $\mathcal{U}'$, respectively, and find

$$\dim(\mathcal{U} + \mathcal{U}') = \dim\mathcal{U} + \dim\mathcal{W}$$

$$\dim\mathcal{U}' = \dim(\mathcal{U} \cap \mathcal{U}') + \dim\mathcal{W}.$$

From these formulas we at once obtain (52.6). ∎

53. Finite dimension and linear mappings

53A. PROPOSITION. *Let the linear space V and the number $n \in \mathbb{N}$ be given. Then V is linearly isomorphic to $\mathbb{F}^{n^{\subset}}$ if and only if V is finite-dimensional and $\dim V = n$.*

Proof. V is finite-dimensional with $\dim V = n$ if and only if there is a finite basis-set $\mathcal{A}$ of V with $\#\mathcal{A} = n$ (definition of "finite-dimensional" and Corollary 52B,(b)). Now a set is finite with cardinal n if and only if it is equinumerous to $n^{\subset}$. The assertion then follows from Proposition 43E ((ii) $\Leftrightarrow$ (iii)). ∎

53B. THEOREM. *Let the linear spaces V and W be given, and assume that one of them is finite-dimensional. Then V is linearly isomorphic to W if and only if both V and W are finite-dimensional and $\dim V = \dim W$.*

Proof. This follows immediately from Proposition 53A. ∎

53C. COROLLARY. *Let the linear space V and the set I be given. The following statements are equivalent.*

(i): *V is finite-dimensional, I is finite, and $\dim V = \#I$.*

(ii): *V is finite-dimensional and linearly isomorphic to $\mathbb{F}^{(I)}$.*

(iii): *V is finite-dimensional and there is a basis of V in V^I.*

(iv): *I is finite and V is linearly isomorphic to $\mathbb{F}^{(I)} = \mathbb{F}^I$.*

(v): *I is finite and there is a basis of V in V^I.*

Proof. In Example 52D,(c) we noted that $\mathbb{F}^{(I)}$ is finite-dimensional if and only if I is finite, and that in that case $\dim \mathbb{F}^{(I)} = \#I$. By Theorem 53B it follows that (i), (ii), (iv) are equivalent. The equivalence of "V is linearly isomorphic to $\mathbb{F}^{(I)}$" and "there is a basis of V in V^I" was noted in Proposition 43E, ((i) $\Leftrightarrow$ (iii)). ∎

53D. PROPOSITION. *Let the linear spaces V and W be given.*

(a): *Assume that W is finite-dimensional. Then there exists an injective linear mapping from V to W if and only if V is finite-dimensional and $\dim V \leq \dim W$.*

(b): *Assume that V is finite-dimensional. Then there exists a surjective linear mapping from V to W if and only if W is finite-dimensional and $\dim W \leq \dim V$.*

Proof. Proof of (a). There exists an injective linear mapping from V to W if and only if V is linearly isomorphic to some subspace of W. Since all subspaces of W are finite-dimensional (Corollary 52F) and their dimensions are precisely the members of $(1 + \dim W)^{\subset}$ (Proposition 52E), it follows from Theorem 53B that it is necessary and sufficient that V be finite-dimensional with $\dim V \leq \dim W$.

Proof of (b). Since V is finite-dimensional, every subspace of V has a supplement in V (Corollary 45G, with Remark 52H). If $L \in \mathrm{Lin}(V, W)$ is surjective, we may choose by Corollary 21I,(b), a linear right-inverse of L, and this is injective. This requires, by (a), that W be finite-dimensional with $\dim W \leq \dim V$. Conversely, if W is finite-dimensional with $\dim W \leq \dim V$, then we may choose, by (a), an injective $M \in \mathrm{Lin}(W, V)$. By Corollary 21I,(a), we may choose a linear left-inverse of M, and this is surjective. ∎

53E. REMARKS. (a): An alternative proof of the "only if" parts of Proposition 53D can be obtained from Theorem 53F, which follows. An alternative proof of the "if" parts of Proposition 53D follows from Proposition 53A and the following observation. Let $m, n \in \mathbb{N}$ be given, with $m \leq n$. Define the linear mappings $H : \mathbb{F}^{m^\sqsubset} \to \mathbb{F}^{n^\sqsubset}$ and $K : \mathbb{F}^{n^\sqsubset} \to \mathbb{F}^{m^\sqsubset}$ by requiring

$$(Ha)|_{m^\sqsubset} = a \quad \text{and} \quad (Ha)|_{n^\sqsubset \setminus m^\sqsubset} = 0 \quad \text{for all } a \in \mathbb{F}^{m^\sqsubset}$$

$$Kb = b|_{m^\sqsubset} \qquad\qquad \text{for all } b \in \mathbb{F}^{n^\sqsubset}.$$

Then $KH = 1_{\mathbb{F}^{m^\sqsubset}}$; therefore H is injective and K is surjective.

(b): Let the linear spaces $\mathcal{V}$ and $\mathcal{W}$ be given. If $\mathcal{V}$ is finite-dimensional and $\mathcal{W}$ is infinite-dimensional, there is an injective linear mapping from $\mathcal{V}$ to $\mathcal{W}$. Indeed we may choose a finite-dimensional subspace $\mathcal{U}$ of $\mathcal{W}$ with $\dim\mathcal{U} = \dim\mathcal{V}$ (Proposition 52E), and further choose a linear isomorphism $L : \mathcal{V} \to \mathcal{U}$ (Theorem 53B). The linear mapping $L|^{\mathcal{W}} : \mathcal{V} \to \mathcal{W}$ is injective.

•If $\mathcal{V}$ is infinite-dimensional and $\mathcal{W}$ is finite-dimensional, there is a surjective linear mapping from $\mathcal{V}$ to $\mathcal{W}$. Indeed, the preceding paragraph shows us that we may choose an injective linear mapping $M : \mathcal{W} \to \mathcal{V}$. By •Corollary 21K,(a), we may choose a linear left-inverse of M, and this is surjective. ∎

We address the question of the finite-dimensionality of product-spaces and co-product spaces. The following result is essentially a Corollary of Theorem 52J.

53F. THEOREM. *Let the family of linear spaces $(\mathcal{V}_i \mid i \in I)$ be given. Let $\mathcal{S}$ be the coproduct-space of a linear coproduct of this family, and $\mathcal{P}$ the product-space of a linear product of this family. The following statements are equivalent.*

(i): *$\operatorname{Supp}(\mathcal{V}_i \mid i \in I)$ is finite and $\mathcal{V}_i$ is finite-dimensional for all $i \in I$.*

(ii): *$\mathcal{S}$ is finite-dimensional.*

(iii): *$\mathcal{P}$ is finite-dimensional.*

If these equivalent statements hold, then

$$\dim\mathcal{S} = \dim\mathcal{P} = \sum_{i \in I} \dim\mathcal{V}_i.$$

Proof. Let $(s_i \mid i \in I)$ be the family of insertions of the linear coproduct in question. For every $i \in I$, s_i is injective (Proposition 32E), hence $\mathcal{V}_i$ and $\operatorname{Rng}s_i$ are linearly isomorphic, and each is finite-dimensional if and only if the other is, with $\dim\mathcal{V}_i = \dim\operatorname{Rng}s_i$ (Theorem 54B). By Corollary 33H, $(\operatorname{Rng}s_i \mid i \in I)$ is a decomposition of $\mathcal{S}$. It therefore follows from Theorem 52J that (i) and (ii) are equivalent and that, if these statements hold, $\dim\mathcal{S} = \sum_{i \in I}\dim\operatorname{Rng}s_i = \sum_{i \in I}\dim\mathcal{V}_i$.

Assume that (i) and (ii) hold. It follows from Proposition 32H that $\mathcal{S}$ and $\mathcal{P}$ are linearly isomorphic, and (iii) follows by Theorem 54B, as does $\dim\mathcal{S} = \dim\mathcal{P}$. Assume, conversely, that (iii) holds; by Proposition 32H there exists an injective linear mapping from $\mathcal{S}$ to $\mathcal{P}$; by Proposition 53D,(a) we conclude that (ii) holds. ∎

53G. EXAMPLES. (a): Let the linear spaces $\mathcal{U}$ and $\mathcal{V}$ be given. Then the linear space $\mathcal{U} \times \mathcal{V}$ is finite-dimensional if and only if $\mathcal{U}$ and $\mathcal{V}$ are finite-dimensional; and in that case $\dim(\mathcal{U} \times \mathcal{V}) = \dim\mathcal{U} + \dim\mathcal{V}$.

(b): Let the non-empty set I and the linear space $\mathcal{V}$ be given. Then the linear space $\mathcal{V}^I$ is finite-dimensional if and only if I is finite and $\mathcal{V}$ is finite dimensional; and in that case $\dim\mathcal{V}^I = (\#I)(\dim\mathcal{V})$. ∎

53H. THEOREM. *Let the linear mapping L be given. Then $\mathrm{Dom}L$ is finite-dimensional if and only if both $\mathrm{Null}L$ and $\mathrm{Rng}L$ are finite-dimensional. In that case,*

$$(53.1) \qquad \dim \mathrm{Dom}L = \dim \mathrm{Null}L + \dim \mathrm{Rng}L.$$

Proof. Proof of the "only if" part. We assume that $\mathrm{Dom}L$ is finite-dimensional. By Corollary 15G (cf. Remark 52H), we may choose a supplement $\mathcal{U}$ of $\mathrm{Null}L$ in $\mathrm{Dom}L$. By Corollary 52F, $\mathrm{Null}L$ and $\mathcal{U}$ are finite-dimensional. By Corollary 52L we have

$$(53.2) \qquad \dim \mathrm{Dom}L = \dim \mathrm{Null}L + \dim\mathcal{U}.$$

By Proposition 21C, the linear mapping $L|_{\mathcal{U}}^{\mathrm{Rng}L}$ is bijective, i.e., a linear isomorphism. By Theorem 53B we infer that $\mathrm{Rng}L$ is finite-dimensional and that $\dim\mathcal{U} = {} = \dim \mathrm{Rng}L$. Combining this formula with (53.2) we obtain (53.1).

Proof of the "if" part. We assume that $\mathrm{Null}L$ and $\mathrm{Rng}L$ are finite-dimensional. Let the finite-dimensional subspace $\mathcal{U}$ of $\mathrm{Dom}L$ be given. We have $\mathrm{Null}(L|_{\mathcal{U}}) = {} = \mathcal{U} \cap \mathrm{Null}L \subset \mathrm{Null}L$ and $\mathrm{Rng}(L|_{\mathcal{U}}) = L_{>}(\mathcal{U}) \subset \mathrm{Rng}L$. We apply the "only if" part of the present theorem to $L|_{\mathcal{U}}$ instead of L and find, using Corollary 52F, that

$$\dim\mathcal{U} = \dim \mathrm{Null}(L|_{\mathcal{U}}) + \dim \mathrm{Rng}(L|_{\mathcal{U}}) \leq \dim \mathrm{Null}L + \dim \mathrm{Rng}L.$$

Since $\mathcal{U}$ was an arbitrary finite-dimensional subspace of $\mathrm{Dom}L$, it follows from Proposition 52E that $\mathrm{Dim}(\mathrm{Dom}L)$ is bounded and that therefore $\mathrm{Dom}L$ is finite-dimensional. ∎

53I. COROLLARY. *Let the linear space $\mathcal{V}$ and the subspace $\mathcal{U}$ of $\mathcal{V}$ be given. Then $\mathcal{V}$ is finite-dimensional if and only if $\mathcal{U}$ and $\mathcal{V}/\mathcal{U}$ are finite-dimensional. In that case,*

$$\dim\mathcal{V} = \dim\mathcal{U} + \dim(\mathcal{V}/\mathcal{U}).$$

Proof. We apply Theorem 53H to $L := \Omega_{\mathcal{V}/\mathcal{U}} \in \mathrm{Lin}(\mathcal{V}, \mathcal{V}/\mathcal{U})$, observing that $\mathrm{Dom}\Omega_{\mathcal{V}/\mathcal{U}} = \mathcal{V}$, $\mathrm{Null}\Omega_{\mathcal{V}/\mathcal{U}} = \mathcal{U}$, $\mathrm{Rng}\Omega_{\mathcal{V}/\mathcal{U}} = \mathcal{V}/\mathcal{U}$ (Section 14). ∎

The next result is reminiscent of the Pigeonhole Principle for mappings from a finite set to a finite set with the same cardinal (*Basic Language*, Corollary 101H).

53J. COROLLARY. *Let the linear mapping L be given. Assume that $\mathrm{Dom}L$ and $\mathrm{Cod}L$ are finite-dimensional. The following statements are equivalent.*

(i): *L is injective and $\dim \mathrm{Dom}L = \dim \mathrm{Cod}L$.*

(ii): L *is surjective and* $\dim \mathrm{Dom}L = \dim \mathrm{Cod}L$.

(iii): L *is invertible.*

Proof. By Theorem 53B, (iii) is equivalent to ((i) and (ii)). It remains to prove that (i) is equivalent to (ii). We may stipulate that

(53.3) $\dim \mathrm{Dom}L = \dim \mathrm{Cod}L$.

By Example 52D,(a), Theorem 53H, (53.3), and Corollary 52F we have the chain of equivalences

$$L \text{ is injective } \Leftrightarrow \quad \mathrm{Null}L = \{0\} \quad \Leftrightarrow \quad \dim \mathrm{Null}L = 0 \quad \Leftrightarrow$$

$$\Leftrightarrow \quad \dim \mathrm{Rng}L = \dim \mathrm{Dom}L \quad \Leftrightarrow \quad \dim \mathrm{Rng}L = \dim \mathrm{Cod}L \quad \Leftrightarrow$$

$$\Leftrightarrow \quad \mathrm{Rng}L = \mathrm{Cod}L \quad \Leftrightarrow \quad L \text{ is surjective. } \blacksquare$$

53K. COROLLARY. *Let the finite-dimensional linear space $\mathcal{V}$ be given. All injective and all surjective members of $\mathrm{Lin}\mathcal{V}$ are invertible.*

The assertion of Corollary 53K actually characterizes finite-dimensional linear spaces, as we now show.

•53L. PROPOSITION. *Let the linear space $\mathcal{V}$ be given. The following statements are equivalent.*

(i): $\mathcal{V}$ *is finite-dimensional.*

(ii): *Every injective member of* $\mathrm{Lin}\mathcal{V}$ *is invertible.*

(iii): *Every surjective member of* $\mathrm{Lin}\mathcal{V}$ *is invertible.*

Proof. (i) *implies* (ii). This follows from Corollary 53K.

•(ii) *implies* (iii). Let a surjective $L \in \mathrm{Lin}\mathcal{V}$ be given. By •Corollary 21K,(b), we may choose a linear right-inverse $M \in \mathrm{Lin}\mathcal{V}$ of L. Then M is injective; by (ii), M is invertible. But L is a left-inverse of M; therefore L is the inverse of M, and L is invertible.

•(iii) *implies* (i). By •Corollary 43M we may choose a basis-set $\mathcal{A}$ of $\mathcal{V}$. Let a surjection $f : \mathcal{A} \to \mathcal{A}$ be given. By Corollary 43I there is exactly one $L \in \mathrm{Lin}\mathcal{V}$ such that $L|_{\mathcal{A}} = f|^{\mathcal{V}}$. Then $L_>(\mathcal{A}) = \mathrm{Rng}f = \mathcal{A}$, and therefore

$$\mathcal{V} = \mathrm{Lsp}\mathcal{A} = \mathrm{Lsp}L_>(\mathcal{A}) \subset \mathrm{Rng}L \subset \mathcal{V},$$

so that L is surjective. By the assumption (iii), L is invertible, hence injective. Therefore $f = L|_{\mathcal{A}}^{\mathcal{A}}$ is injective.

We have shown that every surjection $f : \mathcal{A} \to \mathcal{A}$ is in fact injective. It follows from *Basic Language,* •Corollary 123C that $\mathcal{A}$ is finite. Therefore $\mathcal{V}$ is finite-dimensional.∎

53M. EXAMPLES. (a): Define $S, T \in \text{Lin}\mathbb{F}^{(\mathbb{N})}$ by the rules

$$(Sa)_n := a_{n+1} \quad \text{for all } n \in \mathbb{N} \text{ and } a \in \mathbb{F}^{(\mathbb{N})}$$

$$(Ta)_n := \begin{cases} 0 & \text{if } n = 0 \\ a_{n-1} & \text{if } n \in \mathbb{N}^\times \end{cases} \quad \text{for all } a \in \mathbb{F}^{(\mathbb{N})}$$

Then $ST = 1_{\mathbb{F}^{(\mathbb{N})}}$, so that S is surjective and T is injective; but neither is invertible, since $TS\delta_0^{\mathbb{N}} = 0$. The same formulas define a surjective and injective member of $\text{Lin}\mathbb{F}^{\mathbb{N}}$, neither of which is invertible.

(b)*: With $J \in \text{Lin}(\text{Cont}(\mathbb{R}, \mathbb{R}), \text{Map}(\mathbb{R}, \mathbb{R}))$ as defined in Example 21J, i.e., by the rule

$$(Jf)(t) = \int_0^t f \quad \text{for all } t \in \mathbb{R} \text{ and } f \in \text{Cont}(\mathbb{R}, \mathbb{R}),$$

the linear mapping $J|^{\text{Cont}(\mathbb{R},\mathbb{R})} \in \text{Lin Cont}(\mathbb{R}, \mathbb{R})$ is injective, but not surjective. ∎

Let the finite-dimensional linear spaces $\mathcal{V}$ and $\mathcal{W}$ be given. Let $b \in \mathcal{V}^I$ and $c \in \mathcal{W}^J$ be (finite) bases of $\mathcal{V}$ and $\mathcal{W}$, respectively. For every $L \in \text{Lin}(\mathcal{V}, \mathcal{W})$, we have the linear mapping $(\text{lc}_c^{\mathcal{W}})^{-1} L \text{lc}_b^{\mathcal{V}} \in \text{Lin}(\mathbb{F}^I, \mathbb{F}^J)$ and – denoted by the same symbols – the $J \times I$-matrix corresponding to it under the identification of $\text{Lin}(\mathbb{F}^I, \mathbb{F}^J)$ with $\mathbb{F}^{J \times I}$ described in Section 51. This $J \times I$-matrix is called the **matrix of L with respect to** the bases b, c, and we shall denote it for brevity, by $[L]^{c,b}$. (Several ingenious notational schemes have been proposed for the recording of the bases and the location of the matrix indices, in order to facilitate bookkeeping with these matrices.) We obtain the following formulas from (51.4):

$$(53.4) \qquad Lb_i = \text{lc}_c^{\mathcal{W}}(\text{lc}_c^{\mathcal{W}})^{-1} L \text{lc}_b^{\mathcal{V}} \delta_i^I = \sum_{j \in J} [L]_{j,i}^{c,b} \text{lc}_c^{\mathcal{W}} \delta_j^J = \sum_{j \in J} [L]_{j,i}^{c,b} c_i \quad \text{for all } i \in I,$$

$$(53.5) \quad L\sum_{i \in I} a_i b_i = \sum_{i \in I} a_i Lb_i = \sum_{i \in I} a_i \sum_{j \in J} [L]_{j,i}^{c,b} c_j = \sum_{j \in J} (\sum_{i \in I} [L]_{j,i}^{c,b} a_i) c_j \text{ for all } a \in \mathbb{F}^I.$$

53N. PROPOSITION. *Let the finite-dimensional linear spaces $\mathcal{V}$ and $\mathcal{W}$ and the respective bases $b \in \mathcal{V}^I$ and $c \in \mathcal{W}^J$ be given. The mapping $(L \mapsto [L]^{c,b}) : \text{Lin}(\mathcal{V}, \mathcal{W}) \to \mathbb{F}^{J \times I}$ is a linear isomorphism.*

Proof. In view of the natural linear isomorphism that serves to identify $\text{Lin}(\mathbb{F}^I, \mathbb{F}^J)$ with $\mathbb{F}^{J \times I}$ (Proposition 51A), it suffices to observe that the mapping $(L \mapsto (\text{lc}_c^{\mathcal{W}})^{-1} L \text{lc}_b^{\mathcal{V}}) : \text{Lin}(\mathcal{V}, \mathcal{W}) \to \text{Lin}(\mathbb{F}^I, \mathbb{F}^J)$ is linear and that $(M \mapsto \text{lc}_c^{\mathcal{W}} M (\text{lc}_b^{\mathcal{V}})^{-1}) : \text{Lin}(\mathbb{F}^I, \mathbb{F}^J) \to \text{Lin}(\mathcal{V}, \mathcal{W})$ is its inverse. ∎

53O. PROPOSITION. *Let the finite-dimensional linear spaces $\mathcal{V}$ and $\mathcal{W}$ be given. Then $\text{Lin}(\mathcal{V}, \mathcal{W})$ is finite-dimensional, and*

$$\dim \mathrm{Lin}(\mathcal{V},\mathcal{W}) = (\dim\mathcal{V})(\dim\mathcal{W}).$$

Proof. By Corollary 53C,$((i) \Rightarrow (v))$ we may choose (finite) bases $b \in \mathcal{V}^I$ and $c \in \mathcal{W}^J$ of $\mathcal{V}$ and $\mathcal{W}$, respectively, and find $\#I = \dim\mathcal{V}$, $\#J = \dim\mathcal{W}$. By Proposition 53N, $\mathrm{Lin}(\mathcal{V},\mathcal{W})$ is linearly isomorphic to $\mathbb{F}^{J\times I}$. By Corollary 53C,$((iv) \Rightarrow (i))$, $\mathrm{Lin}(\mathcal{V},\mathcal{W})$ is finite-dimensional and $\dim \mathrm{Lin}(\mathcal{V},\mathcal{W}) = \#(J \times I) = (\#I)(\#J) = (\dim\mathcal{V})(\dim\mathcal{W})$. ∎

53P. PROPOSITION. *Let the finite-dimensional linear spaces $\mathcal{V}$, $\mathcal{W}$, $\mathcal{X}$, and respective bases $b \in \mathcal{V}^I$, $c \in \mathcal{W}^J$, $d \in \mathcal{X}^K$ be given. For all $L \in \mathrm{Lin}(\mathcal{V},\mathcal{W})$ and $M \in \mathrm{Lin}(\mathcal{W},\mathcal{X})$ we have*

$$(53.6) \qquad [ML]^{d,b} = [M]^{d,c}[L]^{c,b},$$

with the multiplication of matrices given by Proposition 51B.

Proof. $(\mathrm{lc}_d^{\mathcal{X}})^{-1}ML\mathrm{lc}_b^{\mathcal{V}} = (\mathrm{lc}_d^{\mathcal{X}})^{-1}M\mathrm{lc}_c^{\mathcal{W}}(\mathrm{lc}_c^{\mathcal{W}})^{-1}L\mathrm{lc}_b^{\mathcal{V}}.$ ∎

53Q. COROLLARY. *Let the finite-dimensional linear spaces $\mathcal{V}$ and $\mathcal{W}$, the bases b and b' of $\mathcal{V}$, and the bases c and c' of $\mathcal{W}$ be given. Then*

$$(53.7) \qquad [L]^{c',b'} = [1_\mathcal{W}]^{c',c}[L]^{c,b}[1_\mathcal{V}]^{b,b'} = ([1_\mathcal{W}]^{c,c'})^{-1}[L]^{c,b}[1_\mathcal{V}]^{b,b'}$$
$$\text{for all } L \in \mathrm{Lin}(\mathcal{V},\mathcal{W}).$$

The matrices $[1_\mathcal{V}]^{b,b'}$ and $[1_\mathcal{W}]^{c,c'}$ are called **change-of-basis matrices** because of their role in (53.7). They express the terms of the "new" basis as linear combinations of the "old": indeed, if $b \in \mathcal{V}^I$, $b' \in \mathcal{V}^{I'}$, (53.4) yields

$$b'_{i'} = 1_\mathcal{V}b'_{i'} = \sum_{i\in I}[1_\mathcal{V}]_{i,i'}^{b,b'}b_i \quad \text{for all } i' \in I'.$$

Let the finite-dimensional linear space $\mathcal{V}$ be given. For every basis b of $\mathcal{V}$ and every $L \in \mathrm{Lin}\mathcal{V}$ we abbreviate the notation by defining $[L]^b := [L]^{b,b}$, and call this square matrix the **matrix of L with respect to b**. In particular, if $b \in \mathcal{V}^I$, we have $[1_\mathcal{V}]^b = \delta^I$, the Kronecker matrix. For $L, M \in \mathrm{Lin}\mathcal{V}$ and bases b and b' of $\mathcal{V}$, formulas (53.6) and (53.7) become, respectively,

$$(53.8) \qquad [ML]^b = [M]^b[L]^b$$

$$(53.9) \qquad [L]^{b'} = ([1_\mathcal{V}]^{b,b'})^{-1}[L]^b[1_\mathcal{V}]^{b,b'}.$$

▼

We combine part of Corollary 53O with a converse.

●**53R.** PROPOSITION. *Let the linear spaces V and W be given. Then $\mathrm{Lin}(V, W)$ is finite-dimensional if and ●only if either both V and W are finite-dimensional, or V is a zero-space, or W is a zero-space.*

Proof. If V or W is a zero-space, $\mathrm{Lin}(V, W)$ is obviously a zero-space. The "if" part therefore follows from Corollary 53O.

To prove the "only if" part, we assume that $\mathrm{Lin}(V, W)$ is finite-dimensional. Let finite-dimensional subspaces U of V and X of W be given. By ●Corollary 21L, the linear mapping $(L \mapsto L|_U) : \mathrm{Lin}(V, W) \to \mathrm{Lin}(U, W)$ is surjective. By Proposition 53D,(b), $\mathrm{Lin}(U, W)$ is finite-dimensional, and

$$(53.10) \qquad \qquad \dim \mathrm{Lin}(U, W) \leq \dim \mathrm{Lin}(V, W).$$

The mapping $(L \mapsto L|^W) : \mathrm{Lin}(U, X) \to \mathrm{Lin}(U, W)$ is obviously linear and injective. By Proposition 53D,(a), $\mathrm{Lin}(U, X)$ is finite-dimensional (cf. Corollary 53O) and

$$(53.11) \qquad \qquad \dim \mathrm{Lin}(U, X) \leq \dim \mathrm{Lin}(U, W).$$

Combining (53.10) and (53.11) with Corollary 53O, we find

$$(53.12) \qquad (\dim U)(\dim X) = \dim \mathrm{Lin}(U, X) \leq \dim \mathrm{Lin}(V, W).$$

This holds for all finite-dimensional subspaces U of V and X of W. If neither V nor W is a zero-space, we may choose $v \in V^\times$ and $w \in W^\times$; then $\dim(\mathbb{F}v) = 1 = \dim(\mathbb{F}w)$, and (53.12) yields

$$\dim U = (\dim U)(\dim(\mathbb{F}w)) \leq \dim \mathrm{Lin}(V, W)$$
$$\text{for all finite-dimensional } U \in \mathrm{Subsp}(V)$$

$$\dim X = (\dim(\mathbb{F}v))(\dim X) \leq \dim \mathrm{Lin}(V, W)$$
$$\text{for all finite-dimensional } X \in \mathrm{Subsp}(W).$$

▲   By Proposition 52E, V and W are finite-dimensional. ∎

▼ 54. Finite- and infinite-dimensional spaces

We have seen that if $\mathcal{V}$ is a finite-dimensional linear space all its basis-sets are equinumerous (Corollary 52B,(b)) and that finite-dimensional linear spaces $\mathcal{V}$ and $\mathcal{W}$ are linearly isomorphic if and only if they have the same dimension (Theorem 53B), i.e., if every basis-set of $\mathcal{V}$ is equinumerous to every basis-set of $\mathcal{W}$. We shall now see that results like these are valid for all linear spaces. There is, however, a sharp distinction in method of proof between the finite-dimensional and the infinite-dimensional cases.

54A. LEMMA. *Let the sets I and I' and the injection $\omega : I' \to I$ be given. For every $a \in \mathbb{F}^{(I)}$ we have $a \circ \omega \in \mathbb{F}^{(I')}$; the mappings $(a \mapsto a \circ \omega) : \mathbb{F}^{(I)} \to \mathbb{F}^{(I')}$ and $(a' \mapsto \sum_{i' \in I'} a'_{i'} \delta^I_{\omega(i')}) : \mathbb{F}^{(I')} \to \mathbb{F}^{(I)}$ are linear, and the latter is a linear right-inverse of the former. If ω is bijective, the latter mapping is $(a' \mapsto a' \circ \omega^{\leftarrow}) : \mathbb{F}^{(I')} \to \mathbb{F}^{(I)}$, and is the inverse of the former.*

Proof. It suffices to observe that $\mathrm{Supp}(a \circ \omega) = \omega^{<}(\mathrm{Supp}\, a)$ for all $a \in \mathbb{F}^{(I)}$, and that $\delta^I_{\omega(i')} \circ \omega = \delta^{I'}_{i'}$ for all $i' \in I'$. ∎

•**54B. THEOREM.** *Let the sets I and I' be given. The following statements are equivalent.*

(i): *I outnumbers I'.*

(ii): *There is a linearly right-invertible linear mapping from $\mathbb{F}^{(I)}$ to $\mathbb{F}^{(I')}$.*

(iii): *There is a linearly left-invertible linear mapping from $\mathbb{F}^{(I')}$ to $\mathbb{F}^{(I)}$.*

(iv): *There is a surjective linear mapping from $\mathbb{F}^{(I)}$ to $\mathbb{F}^{(I')}$.*

(v): *There is an injective linear mapping from $\mathbb{F}^{(I')}$ to $\mathbb{F}^{(I)}$.*

Proof. (i) *implies* (iii). This follows from Lemma 54A.

(iii) *implies* (ii). This is trivial.

(ii) *implies* (iv). This is trivial (cf. Corollary 21K,(b)).

•(iii) *is equivalent to* (v). This follows from •Corollary 21K,(a).

•(iv) *implies* (i). We assume that there is a surjective linear mapping from $\mathbb{F}^{(I)}$ to $\mathbb{F}^{(I')}$.

We first suppose that I is finite. Then $\mathbb{F}^{(I)}$ is finite-dimensional, with $\dim \mathbb{F}^{(I)} = \#I$ (Example 52D,(c)). By Proposition 53D,(b), $\mathbb{F}^{(I')}$ must also be finite-dimensional, with $\dim \mathbb{F}^{(I')} \le \dim \mathbb{F}^{(I)}$. Then I' is finite, with $\#I' = \dim \mathbb{F}^{(I')}$. Combining these facts, we find that $\#I' \le \#I$, and therefore I outnumbers I' (*Basic Language*, Proposition 101E).

We suppose from now on that I is infinite, and choose a surjective $L \in \mathrm{Lin}(\mathbb{F}^{(I)}, \mathbb{F}^{(I')})$. Let $j' \in I'$ be given, and choose $a \in \mathbb{F}^{(I)}$ such that $La = \delta^{I'}_{j'}$. By (13.4) we have

$$1 = \delta^{I'}_{j',j'} = (La)_{j'} = (L \sum_{i \in I} a_i \delta^I_i)_{j'} = \sum_{i \in I} a_i (L\delta^I_i)_{j'},$$

and therefore there exists $j \in I$ such that $(L\delta^I_j)_{j'} \ne 0$, i.e., $j' \in \mathrm{Supp}(L\delta^I_j)$. Since

$j' \in I'$ was arbitrary, we have shown that

$$I' = \bigcup_{i \in I} \mathrm{Supp}(L\delta_i^I).$$

Since $\mathrm{Supp}(L\delta_i^I)$ is finite, and hence countable, for every $i \in I$, it follows from •Tukey's Lemma ((VII) in *Basic Language*, Section 173) and *Basic Language*, Corollary 175H,(b) that the infinite set I outnumbers I'. ∎

•**54C. COROLLARY.** *Let the sets I and I' be given. Then $\mathbb{F}^{(I')}$ is linearly isomorphic to $\mathbb{F}^{(I)}$ if and •only if I is equinumerous to I'.*

Proof. The "if" part follows from Lemma 54A. To prove the •"only if" part, assume that $\mathbb{F}^{(I)}$ is linearly isomorphic to $\mathbb{F}^{(I')}$. By •Theorem54B ((ii) $\Rightarrow$ (i) and (iii) $\Rightarrow$ (i)) it follows that I outnumbers I' and I' outnumbers I. By the Schröder-Bernstein Theorem (*Basic Language*, Theorem 75C; cf. *Basic Language*, Theorem 175A), I and I' are equinumerous. ∎

•**54D. COROLLARY.** *Let the linear space V be given.*

(a): *If $b \in V^I$ and $b' \in V^{I'}$ are bases of V, then I and I' are equinumerous.*

(b): *If B and B' are basis-sets of V, then B and B' are equinumerous.*

Proof. $(\mathrm{lc}_{b'}^V)^{-1}\mathrm{lc}_b^V \in \mathrm{Lin}(\mathbb{F}^{(I)}, \mathbb{F}^{(I')})$ is a linear isomorphism. By •Corollary 54C, I and I' are equinumerous. This establishes (a); and (b) is a special case of (a). ∎

•**54E. COROLLARY.** *Let the linear spaces V and W, and respective bases $b \in V^I$ and $c \in W^J$ be given.*

(a): *The following statements are equivalent.*

  (i): *I outnumbers J;*
  (ii): *there is a linearly right-invertible linear mapping from V to W.*
  (iii): *there is a linearly left-invertible linear mapping from W to V.*
  (iv): *there is a surjective linear mapping from V to W.*
  (v): *there is an injective linear mapping from W to V.*

(b): *V is linearly isomorphic to W if and only if I and J are equinumerous.*

Proof. V is linearly isomorphic to $\mathbb{F}^{(I)}$, and W is linearly isomorphic to $\mathbb{F}^{(J)}$. The assertion now follows from •Theorem 54B and •Corollary 54C. ∎

54F. REMARK. In *Basic Language*, Remark 175D,(a), we mentioned the possibility of generalizing the notion of cardinal of a finite set by assigning a *cardinal $\#S$* to *every* set S in such a way that, for given sets S and T, we have $\#S = \#T$ if and only if S and T are equinumerous. We also mentioned that one could define a "relation" $\leq$ among the *cardinal numbers* that occur as cardinals of sets, in such a way that $\#S \leq \#T$ if and only if T outnumbers S. By virtue of •Corollary 54C we may then define $\dim V$, the *dimension of* a linear space V, as the cardinal $\#B$ of every basis-set B of V (the existence of a basis-set is assured by •Corollary 43M). Corollary 52B,(b) shows that this definition does not clash with the definition (52.1) of the dimension of a finite-dimensional linear space. It follows from •Corollary 54E,(b) that, for given linear spaces V and W, V is linearly isomorphic to W if and only if $\dim V = \dim W$; and from •Corollary 54E,(a) that statements (ii)-(v) in that corollary are equivalent to $\dim W \leq \dim V$. ∎

Chapter 6

DUALITY

61. Dual spaces and transposes

We begin by recalling a definition, given in Examples 12G,(g). Let the linear space $\mathcal{V}$ be given. For every $u \in \mathcal{V}$ we define the linear mapping $u\otimes \in \mathrm{Lin}(\mathbb{F}, \mathcal{V})$ by the rule

$$u \otimes t := tu \quad \text{for all } t \in \mathbb{F}.$$

When it is necessary to specify the codomain $\mathcal{V}$ of this mapping, we write $u\otimes^{\mathcal{V}}$ in full. However, if $\mathcal{U}$ is a subspace of $\mathcal{V}$ we obviously have

$$u \otimes^{\mathcal{U}} = u \otimes^{\mathcal{V}}|^{\mathcal{U}} \quad \text{for all } u \in \mathcal{U},$$

and therefore the omission of this indication of codomain causes no clashes in so far as the values of these mappings are concerned.

61A. Proposition. *Let the linear space $\mathcal{V}$ be given. The mappings*

$$u \mapsto u\otimes^{\mathcal{V}} : \mathcal{V} \to \mathrm{Lin}(\mathbb{F}, \mathcal{V})$$

$$L \mapsto L1 : \mathrm{Lin}(\mathbb{F}, \mathcal{V}) \to \mathcal{V}$$

are linear, and each is the inverse of the other.

Proof. The linearity of either mapping is obvious. (We note that the latter mapping is $\mathrm{ev}^{\mathrm{Lin}(\mathbb{F},\mathcal{V})}{}_1$.) We have

$$u \otimes 1 = 1u = u \quad \text{for all } u \in \mathcal{V}.$$

On the other hand,

$$(L1) \otimes t = tL1 = L(t1) = Lt \quad \text{for all } t \in \mathbb{F} \text{ and } L \in \mathrm{Lin}(\mathbb{F}, \mathcal{V}),$$

so that $(L1)\otimes^{\mathcal{V}} = L$ for all $L \in \mathrm{Lin}(\mathbb{F}, \mathcal{V})$. ∎

61B. REMARK. Proposition 61A shows that $(u \mapsto u \otimes^{\mathcal{V}}) : \mathcal{V} \to \mathrm{Lin}(\mathbb{F}, \mathcal{V})$ is a linear isomorphism. It might be used to *identify* these linear spaces, i.e., to write u instead of $u \otimes^{\mathcal{V}}$ or $u \otimes$ for every $u \in \mathcal{V}$. We prefer not to do this, however, since it would lead to notational clashes later. There is one exception: when $\mathcal{V}$ is the linear space $\mathbb{F}$ (Examples 11D,(b)), we do identify $\mathbb{F}$ and $\mathrm{Lin}\mathbb{F}$ in this way, and write s instead of $s \otimes^{\mathbb{F}}$ or $s \otimes$ for every $s \in \mathbb{F}$. In view of the commutativity of multiplication in $\mathbb{F}$, this produces no clash: indeed,

$$s \otimes t = ts = st \quad \text{for all } s, t \in \mathbb{F}.$$

We observe that, under this identification, $s = \mathrm{mult}^{\mathbb{F}}_s$ for all $s \in \mathbb{F}$. This identification induces others: e.g., of $\mathrm{Lin}(\mathbb{F}, \mathcal{V})$ with $\mathrm{Lin}(\mathrm{Lin}\mathbb{F}, \mathcal{V})$ and of $\mathrm{Lin}(\mathcal{V}, \mathbb{F})$ with $\mathrm{Lin}(\mathcal{V}, \mathrm{Lin}\mathbb{F})$ for every linear space $\mathcal{V}$. ∎

For each given linear space $\mathcal{V}$, we define the linear space $\mathcal{V}^* := \mathrm{Lin}(\mathcal{V}, \mathbb{F})$, and call $\mathcal{V}^*$ the **dual space of** $\mathcal{V}$. The members of $\mathcal{V}^*$ are called **linear forms on** $\mathcal{V}$ or *linear functionals on* $\mathcal{V}$. We observe that $\mathbb{F}^* = \mathrm{Lin}\mathbb{F}$ is identified with $\mathbb{F}$ in the sense of Remark 61B. We also note that the dual space of a zero-space is a zero-space.

Let the linear spaces $\mathcal{V}$ and $\mathcal{W}$ be given. For each linear mapping $L \in \mathrm{Lin}(\mathcal{V}, \mathcal{W})$, the mapping

$$\mu \mapsto \mu L : \mathcal{W}^* \to \mathcal{V}^*$$

is linear (Remark 13I); we denote this linear mapping by $L^{\top}$ and call it the **transpose of** L. (Some mathematicians denote it by L^*, and call it the *adjoint of* L.) We thus have

(61.1) $L^{\top} \mu = \mu L \quad \text{for all } \mu \in \mathcal{W}^* \text{ and } L \in \mathrm{Lin}(\mathcal{V}, \mathcal{W}).$

We summarize some elementary properties of the transposes of linear mappings.

61C. PROPOSITION. (a): *Let the linear spaces $\mathcal{V}$ and $\mathcal{W}$ be given. The mapping* $L \mapsto L^{\top} : \mathrm{Lin}(\mathcal{V}, \mathcal{W}) \to \mathrm{Lin}(\mathcal{W}^*, \mathcal{V}^*)$ *is linear.*

(b): *Let the linear spaces $\mathcal{V}, \mathcal{W}, \mathcal{X}$ be given. Then $(ML)^{\top} = L^{\top} M^{\top}$ for all*

$L \in \mathrm{Lin}(\mathcal{V}, \mathcal{W})$ *and* $M \in \mathrm{Lin}(\mathcal{W}, \mathcal{X})$.

(c): *For every linear space $\mathcal{V}$ we have $1_{\mathcal{V}}{}^{\top} = 1_{\mathcal{V}^*}$.*

(d): *Let the linear spaces $\mathcal{V}$ and $\mathcal{W}$ be given. For every $L \in \mathrm{Lis}(\mathcal{V}, \mathcal{W})$, we have* $L^{\top} \in \mathrm{Lis}(\mathcal{W}^*, \mathcal{V}^*)$ *and* $(L^{\top})^{-1} = (L^{-1})^{\top}$.

Proof. Proof of (d). Let $L \in \mathrm{Lis}(\mathcal{V}, \mathcal{W})$ be given. By (b) and (c) we have $L^{\top}(L^{-1})^{\top} = (L^{-1}L)^{\top} = 1_{\mathcal{V}}{}^{\top} = 1_{\mathcal{V}^*}$ and $(L^{-1})^{\top} L^{\top} = (LL^{-1})^{\top} = 1_{\mathcal{W}}{}^{\top} = 1_{\mathcal{W}^*}$. ∎

The linear mapping $(L \mapsto L^{\top}) : \mathrm{Lin}(\mathcal{V}, \mathcal{W}) \to \mathrm{Lin}(\mathcal{W}^*, \mathcal{V}^*)$ is called the **linear transposition (for** the pair of linear spaces $(\mathcal{V}, \mathcal{W})$).

61D. EXAMPLES. (a): Let the linear space $\mathcal{V}$ be given. For every $\lambda \in \mathcal{V}^*$ we have $\lambda^{\top} \in \mathrm{Lin}(\mathbb{F}^*, \mathcal{V}^*)$; under the identification of $\mathrm{Lin}(\mathbb{F}^*, \mathcal{V}^*)$ with $\mathrm{Lin}(\mathbb{F}, \mathcal{V}^*)$ (Remark 61B) we have

$$(\lambda^\top t)v = t(\lambda v) = (t\lambda)v = (\lambda \otimes t)v \quad \text{for all } t \in \mathbb{F} \text{ and } v \in \mathcal{V};$$

thus

(62.2) $$\lambda^\top = \lambda \otimes \quad \text{for all } \lambda \in \mathcal{V}^*.$$

●(b): Let the linear space $\mathcal{V}$ and the subspace $\mathcal{U}$ of $\mathcal{V}$ be given. Then the linear mapping $1_{\mathcal{U} \subset \mathcal{V}}{}^\top : \mathcal{V}^* \to \mathcal{U}^*$ is the operation of restricting a linear form on $\mathcal{V}$ to $\mathcal{U}$: for all $\lambda \in \mathcal{V}^*$ we have, by (61.1), $1_{\mathcal{U} \subset \mathcal{V}}{}^\top \lambda = \lambda 1_{\mathcal{U} \subset \mathcal{V}} = \lambda|_{\mathcal{U}}$. It follows from ●Corollary 21L that $1_{\mathcal{U} \subset \mathcal{V}}{}^\top$ is surjective. ■

61E. Remark. In practice, the notation $u\otimes$ recalled at the beginning of this section occurs most frequently in composition: for given linear spaces $\mathcal{V}$ and $\mathcal{W}$ we have the composite $w \otimes \lambda \in \mathrm{Lin}(\mathcal{V}, \mathcal{W})$ for every $w \in \mathcal{W}$ and $\lambda \in \mathcal{V}^*$; and

(61.3) $$(w \otimes \lambda)v = (\lambda v)w \quad \text{for all } v \in \mathcal{V} \text{ and } \lambda \in \mathcal{V}^*. ■$$

We examine with some care the structure of the dual space of $\mathbb{F}^{(I)}$ for a given set I. For every $a \in \mathbb{F}^I$ we consider the unique linear form $a\cdot \in (\mathbb{F}^{(I)})^*$ that satisfies

(61.4) $$a\cdot\delta_i^I = a_i \quad \text{for all } i \in I$$

(Theorem 43F,(c)). These linear forms satisfy

(61.5) $$a\cdot s = \sum_{i \in I} a_i s_i \quad \text{for all } a \in \mathbb{F}^I \text{ and all } s \in \mathbb{F}^{(I)},$$

as follows from (61.4) by (13.4). In particular, we have

(61.6) $$\delta_j^I \cdot s = s_j \quad \text{for all } j \in I \text{ and } s \in \mathbb{F}^{(I)}.$$

61F. Proposition. *Let the set I be given. Then $\mathbb{F}^I$ is linearly isomorphic to $(\mathbb{F}^{(I)})^*$. More precisely, the mappings*

$$a \mapsto a\cdot : \mathbb{F}^I \to (\mathbb{F}^{(I)})^*$$
$$\lambda \mapsto (\lambda\delta_i^I \mid i \in I) : (\mathbb{F}^{(I)})^* \to \mathbb{F}^I$$

are linear, and each is the inverse of the other.

Proof. The mappings are obviously linear. By (61.4) we have

$$(a\cdot\delta_i^I \mid i \in I) = (a_i \mid i \in I) = a \quad \text{for all } a \in \mathbb{F}^I.$$

On the other hand, (61.4) yields

$$(\lambda \delta_i^I \mid i \in I) \cdot \delta_j^I = \lambda \delta_j^I \quad \text{for all } j \in I \text{ and } \lambda \in (\mathbb{F}^{(I)})^*,$$

and therefore

$$(\lambda \delta_i^I \mid i \in I) \cdot = \lambda \quad \text{for all } \lambda \in (\mathbb{F}^{(I)})^*. \ \blacksquare$$

61G. Remark. We do not, strictly speaking, identify the linear spaces $\mathbb{F}^I$ and $(\mathbb{F}^{(I)})^*$ by means of the linear isomorphism $(a \mapsto a \cdot)$, since we maintain the notational distinction between a and $a \cdot$. However, we use such notations as $K \cdot$ for the image under this isomorphism of a subset K of $\mathbb{F}^I$. $\blacksquare$

•**61H. Corollary.** *For every linear space $\mathcal{V}$ there is an injective linear mapping from $\mathcal{V}$ to $\mathcal{V}^*$.*

Proof. Let the linear space $\mathcal{V}$ be given. By •Corollary 43M, we may choose a set I and a linear isomorphism $L \in \mathrm{Lis}(\mathcal{V}, \mathbb{F}^{(I)})$. Then $L^\top \in \mathrm{Lis}((\mathbb{F}^{(I)})^*, \mathcal{V}^*)$, by Proposition 61C,(d). The composite of the linear isomorphism L, the (linear) inclusion of $\mathbb{F}^{(I)}$ in $\mathbb{F}^I$, the linear isomorphism $(a \mapsto a \cdot) : \mathbb{F}^I \to (\mathbb{F}^{(I)})^*$, and the linear isomorphism $L^\top$ is an injective linear mapping from $\mathcal{V}$ to $\mathcal{V}^*$. $\blacksquare$

62. Annihilators

Let the linear space $\mathcal{V}$ be given. For every subset $\mathcal{A}$ of $\mathcal{V}$ we define

$$\mathcal{A}^{\perp} := \{\lambda \in \mathcal{V}^* \mid \forall v \in \mathcal{A}, \ \lambda v = 0\} = \{\lambda \in \mathcal{V}^* \mid \lambda|_{\mathcal{A}} = 0_{\mathcal{A} \to \mathbb{F}}\} =$$

$$= \{\lambda \in \mathcal{V}^* \mid \lambda_>(\mathcal{A}) \subset \{0\}\} = \{\lambda \in \mathcal{V}^* \mid \mathcal{A} \subset \mathrm{Null}\lambda\},$$

and call the subset of $\mathcal{V}^*$ the **annihilator of** $\mathcal{A}$ (sometimes read "$\mathcal{A}$-**perp**"). For every subset of $\mathcal{B}$ of $\mathcal{V}^*$ we define

$$\mathcal{B}_{\perp} := \{v \in \mathcal{V} \mid \forall \lambda \in \mathcal{B}, \ \lambda v = 0\} = \bigcap_{\lambda \in \mathcal{B}}{}^{\mathcal{V}^*}\mathrm{Null}\lambda,$$

and call this subset of $\mathcal{B}$ the **pre-annihilator of** $\mathcal{B}$.

62A. REMARK. Let the linear space $\mathcal{V}$, the proper subspace $\mathcal{U}$ of $\mathcal{V}$, and the subset $\mathcal{A}$ of $\mathcal{U}$ be given. Then the annihilator of the set $\mathcal{A}$ regarded as a subset of $\mathcal{U}$ is not the same set as the annihilator of the set $\mathcal{A}$ regarded as a subset of $\mathcal{V}$ (these annihilators are subsets of $\mathcal{U}^*$, and of $\mathcal{V}^*$, respectively). If necessary, they are distinguished by calling the former the $\mathcal{U}$-**annihilator of** $\mathcal{A}$ and the latter the $\mathcal{V}$-**annihilator of** $\mathcal{A}$. A generic notational distinction is too cumbersome; appropriate notations will be agreed on as needed in each case. ∎

62B. PROPOSITION. *Let the linear space $\mathcal{V}$ be given.*

(a): *$\mathcal{A}^{\perp}$ is a subspace of $\mathcal{V}^*$ for every subset $\mathcal{A}$ of $\mathcal{V}$, and $\mathcal{B}_{\perp}$ is a subspace of $\mathcal{V}$ for every subset $\mathcal{B}$ of $\mathcal{V}^*$.*

(b): *$\varnothing^{\perp} = \{0\}^{\perp} = \mathcal{V}^*$ and $\mathcal{V}^{\perp} = \{0\}$ and $\varnothing_{\perp} = \{0\}_{\perp} = \mathcal{V}$.*

62C. PROPOSITION. *Let the linear space $\mathcal{V}$ be given.*

(a): *The pair of mappings whose components are*

(62.1) $$\mathcal{A} \mapsto \mathcal{A}^{\perp} : \mathfrak{P}(\mathcal{V}) \to \mathfrak{P}(\mathcal{V}^*)$$

(62.2) $$\mathcal{B} \to \mathcal{B}_{\perp} : \mathfrak{P}(\mathcal{V}^*) \to \mathfrak{P}(\mathcal{V})$$

is the Galois correspondence from $\mathfrak{P}(\mathcal{V})$ to $\mathfrak{P}(\mathcal{V}^*)$ (both ordered by inclusion) associated as in Basic Language, Proposition 74C with the relation $\perp$ from $\mathcal{V}$ to $\mathcal{V}^*$ defined by

$$\forall v \in \mathcal{V}, \forall \lambda \in \mathcal{V}^*, \quad v \perp \lambda \quad :\Leftrightarrow \quad \lambda v = 0.$$

In particular, the mappings (62.1) and (62.2) are antitone.

(b): *The mappings*

(62.3) $$\mathcal{A} \mapsto \mathcal{A}^{\perp}{}_{\perp} : \mathfrak{P}(\mathcal{V}) \to \mathfrak{P}(\mathcal{V})$$

(62.4) $$\mathcal{B} \mapsto \mathcal{B}_\perp{}^\perp : \mathfrak{P}(\mathcal{V}^*) \to \mathfrak{P}(\mathcal{V}^*)$$

are closure mappings in $\mathfrak{P}(\mathcal{V})$ and in $\mathfrak{P}(\mathcal{V}^*)$ (ordered by inclusion), respectively.

(c): $\mathcal{A}^\perp{}_\perp{}^\perp = \mathcal{A}^\perp$ *for every subset $\mathcal{A}$ of $\mathcal{V}$, and $\mathcal{B}_\perp{}^\perp{}_\perp = \mathcal{B}_\perp$ for every subset $\mathcal{B}$ of* $\mathcal{V}^*$.

Proof. (a) follows immediately from the definitions; (b) and (c) follow from (a) by *Basic Language*, Theorem 74E. ∎

62D. COROLLARY. *Let the linear space $\mathcal{V}$ be given.*

(a): $\mathrm{Lsp}\mathcal{A} \subset \mathcal{A}^\perp{}_\perp$ *and $(\mathrm{Lsp}\mathcal{A})^\perp = \mathcal{A}^\perp$ for every subset $\mathcal{A}$ of $\mathcal{V}$, and $\mathrm{Lsp}\mathcal{B} \subset \mathcal{B}_\perp{}^\perp$ and $(\mathrm{Lsp}\mathcal{B})_\perp = \mathcal{B}_\perp$ for every subset $\mathcal{B}$ of* $\mathcal{V}^*$.

(b): *For every family $(\mathcal{A}_i \mid i \in I)$ of subsets of $\mathcal{V}$ we have*

(62.5) $$\bigcap_{i \in I}{}^{\mathcal{V}^*} \mathcal{A}_i{}^\perp = \left(\bigcup_{i \in I} \mathcal{A}_i\right)^\perp = \left(\sum_{i \in I} \mathrm{Lsp}\mathcal{A}_i\right)^\perp;$$

and for every family $(\mathcal{B}_i \mid i \in I)$ of subsets of $\mathcal{V}^*$ we have

(62.6) $$\bigcap_{i \in I}{}^{\mathcal{V}} \mathcal{B}_{i\perp} = \left(\bigcup_{i \in I} \mathcal{B}_i\right)_\perp = \left(\sum_{i \in I} \mathrm{Lsp}\mathcal{B}_i\right)_\perp.$$

Proof. Proof of (a). Let the subset of $\mathcal{A}$ of $\mathcal{V}$ be given. By Propositions 62B,(a) and 62C,(b), $\mathcal{A}^\perp{}_\perp$ is a subspace of $\mathcal{V}$ and includes $\mathcal{A}$. Hence $\mathcal{A} \subset \mathrm{Lsp}\mathcal{A} \subset \mathcal{A}^\perp{}_\perp$. From Proposition 62C,(a),(c) we then obtain $\mathcal{A}^\perp \supset (\mathrm{Lsp})^\perp \supset \mathcal{A}^\perp{}_\perp{}^\perp = \mathcal{A}^\perp$, so that equality holds. The proof of the second part of (a) is entirely similar.

Proof of (b). Let the family $(\mathcal{A}_i \mid i \in I)$ of subsets of $\mathcal{V}$ be given. It follows from the definition of annihilator that

$$\bigcap_{i \in I}{}^{\mathcal{V}} \mathcal{A}_i{}^\perp = \left(\bigcup_{i \in I} \mathcal{A}_i\right)^\perp.$$

We now apply the same argument to the family $(\mathrm{Lsp}\mathcal{A}_i \mid i \in I)$, together with (a) (twice) and Proposition 13R, and find

$$\bigcap_{i \in I}{}^{\mathcal{V}^*} \mathcal{A}_i^\perp = \bigcap_{i \in I}{}^{\mathcal{V}^*} (\mathrm{Lsp}\mathcal{A}_i)^\perp = \left(\bigcup_{i \in I} \mathrm{Lsp}\mathcal{A}_i\right)^\perp = \left(\mathrm{Lsp}\bigcup_{i \in I} \mathrm{Lsp}\mathcal{A}_i\right)^\perp = \left(\sum_{i \in I} \mathrm{Lsp}\mathcal{A}_i\right)^\perp.$$

This completes the proof of (62.5). The proof of (62.6) is entirely similar. ∎

62E. COROLLARY. *Let the linear space $\mathcal{V}$ be given.*

(a): *The pair of mappings whose components are*

(62.7) $$\mathcal{U} \mapsto \mathcal{U}^\perp : \mathrm{Subsp}(\mathcal{V}) \to \mathrm{Subsp}(\mathcal{V}^*)$$

(62.8) $W \mapsto W_\perp : \mathrm{Subsp}(V^*) \to \mathrm{Subsp}(V)$

is a Galois correspondence from $\mathrm{Subsp}(V)$ to $\mathrm{Subsp}(V^*)$ (both ordered by inclusion).
   (b): For every family $(U_i \mid i \in I)$ of subspaces of V we have

$$\bigcap_{i \in I}^{V^*} U_i^\perp = \left(\sum_{i \in I} U_i \right)^\perp,$$

and for every family $(W_i \mid i \in I)$ of subspaces of V^* we have

$$\bigcap_{i \in I}^{V^*} W_{i\perp} = \left(\sum_{i \in I} W_i \right)_\perp.$$

62F. Proposition. Let the linear spaces V and W and the linear mapping $L \in \mathrm{Lin}(V, W)$ be given.
   (a): $L_>(A)^\perp = (L^\top)^<(A^\perp)$ for every subset A of V. In particular,

$$(\mathrm{Rng}L)^\perp = \mathrm{Null}L^\top.$$

   (b): If L is surjective, then $L^\top$ is injective.
   (c): $(L^\top)_>(B)_\perp = L^<(B_\perp)$ for every subset B of W^*. In particular,

$$(\mathrm{Rng}L^\top)_\perp = L^<(W^*_\perp).$$

   Proof. Proof of (a). Let the subset A of V be given. For each $\mu \in W^*$ we have the chain of equivalences

$$\mu \in L_>(A)^\perp \Leftrightarrow (\forall v \in A, \ \mu Lv = 0) \Leftrightarrow (\forall v \in A, \ (L^\top \mu)v = 0) \Leftrightarrow$$
$$\Leftrightarrow L^\top \mu \in A^\perp \Leftrightarrow \mu \in (L^\top)^<(A^\perp).$$

The special case is obtained by setting $A := V$ and recalling that $V^\perp = \{0\}$.
   Proof of (c). Let the subset B of W^* be given. For every $v \in V$ we have the chain of equivalences

$$v \in (L^\top)_>(B)_\perp \Leftrightarrow (\forall \mu \in B, \ (L^\top \mu)v = 0) \Leftrightarrow (\forall \mu \in B, \ \mu Lv = 0) \Leftrightarrow$$
$$\Leftrightarrow Lv \in B_\perp \Leftrightarrow v \in L^<(B_\perp). \ \blacksquare$$

   We next obtain an improvement on the result in Corollary 62D,(a) by showing that $A^{\perp}{}_\perp = \mathrm{Lsp}A$ for all subsets A of a linear space ($\bullet$Theorem 62H,(b)) and that $B_\perp{}^\perp = \mathrm{Lsp}B$ for certain subsets B of the dual space (Theorem 62K).
   $\bullet$**62G. Lemma.** Let the linear space V, the subspace U of V, and $v \in V \backslash U$ be given. Then there exists $\lambda \in U^\perp$ such that $\lambda v = 1$.
   Proof. The linear mapping $v \otimes^{\mathbb{F}v}$, i.e., $(t \mapsto tv) : \mathbb{F} \to \mathbb{F}v$ is injective, since $v \neq 0$, and surjective. Let $\sigma \in (\mathbb{F}v)^*$ be its inverse; it satisfies $\sigma v = 1$.

Now $\mathbb{F}v \cap \mathcal{U} = \{0\}$. by •Theorem 15F we may choose a supplement $\mathcal{H}$ of $\mathbb{F}v$ in $\mathcal{V}$ such that $\mathcal{U} \subset \mathcal{H}$. By Proposition 15C we may choose $P \in \mathrm{Lin}(\mathcal{V}, \mathbb{F}v)$ such that $P|_{\mathbb{F}v} = 1_{\mathbb{F}v}$ and $\mathrm{Null}P = \mathcal{H}$. Then $\lambda := \sigma P \in \mathcal{V}^*$ satisfies $\lambda v = \sigma P v = \sigma v = 1$ and $\mathcal{U} \subset \mathcal{H} = \mathrm{Null}P \subset \mathrm{Null}\lambda$, so that $\lambda \in \mathcal{U}^\perp$. ∎

•62H. Theorem. *Let the linear space $\mathcal{V}$ be given.*

(a): $\mathcal{U}^{\perp}{}_\perp = \mathcal{U}$ *for every subspace $\mathcal{U}$ of $\mathcal{V}$; in particular, $\mathcal{V}^*{}_\perp = \{0\}$.*

(b): $\mathcal{A}^{\perp}{}_\perp = \mathrm{Lsp}\mathcal{A}$ *for every subset $\mathcal{A}$ of $\mathcal{V}$.*

(c): *The mapping (62.7) is a right-inverse of the mapping (62.8); in particular, the former mapping is injective and the latter mapping is surjective.*

Proof. Proof of (a)*.* Let the subspace $\mathcal{U}$ of $\mathcal{V}$ be given. Let $v \in \mathcal{U}^{\perp}{}_\perp$ be given. Then $\lambda v = 0$ for all $\lambda \in \mathcal{U}^\perp$. It follows by •Lemma 62G that $v \notin \mathcal{V}\backslash\mathcal{U}$, i.e., that $v \in \mathcal{U}$. Thus $\mathcal{U}^{\perp}{}_\perp \subset \mathcal{U}$. But $\mathcal{U} \subset \mathcal{U}^{\perp}{}_\perp$ by Proposition 62C,(b), so that equality holds. From Proposition 62B,(b) we then have $\mathcal{V}^*{}_\perp = \{0\}^{\perp}{}_\perp = \{0\}$.

(b). By (a) and Corollary 62D,(a) we have $\mathcal{A}^{\perp}{}_\perp = (\mathrm{Lsp}\mathcal{A})^{\perp}{}_\perp = \mathrm{Lsp}\mathcal{A}$ for every subset $\mathcal{A}$ of $\mathcal{V}$.

(c). This is a reformulation of (a). ∎

•62I. Corollary. *Let the linear spaces $\mathcal{V}$ and $\mathcal{W}$ and the linear mapping $L \in \mathrm{Lin}(\mathcal{V}, \mathcal{W})$ be given. Then*

$$(\mathrm{Rng}L^\top)_\perp = \mathrm{Null}L \quad and \quad (\mathrm{Null}L^\top)_\perp = \mathrm{Rng}L.$$

If $L^\top$ is surjective then L is injective; if $L^\top$ is injective then L is surjective.

Proof. By Proposition 62F and •Theorem 62H,(a) we have

$$(\mathrm{Rng}L^\top)_\perp = L^<(\mathcal{W}^*{}_\perp) = L^<(\{0\}) = \mathrm{Null}L$$

$$(\mathrm{Null}L^\top)_\perp = (\mathrm{Rng}L)^{\perp}{}_\perp = \mathrm{Rng}L.$$

If $L^\top$ is surjective, we have $\mathrm{Null}L = \mathcal{V}^*{}_\perp = \{0\}$, so that L is injective. If $L^\top$ is injective, we have $\mathrm{Rng}L = \{0\}_\perp = \mathcal{W}$, so that L is surjective. ∎

•62J. Corollary. *For every pair $(\mathcal{V}, \mathcal{W})$ of linear spaces, the linear transposition is injective.*

Proof. For every $L \in \mathrm{Lin}(\mathcal{V}, \mathcal{W})$ we have, by •Corollary 62I, the chain of implications

$$L^\top = 0 \quad \Rightarrow \quad \mathrm{Rng}L^\top = \{0\} \quad \Rightarrow \quad \mathrm{Null}L = (\mathrm{Rng}L^\top)_\perp = \{0\}_\perp = \mathcal{V} \quad \Rightarrow \quad L = 0.$$

Hence the linear mapping $(L \mapsto L^\top) : \mathrm{Lin}(\mathcal{V}, \mathcal{W}) \to \mathrm{Lin}(\mathcal{W}^*, \mathcal{V}^*)$ is injective. ∎

62K. Theorem. *Let the linear space $\mathcal{V}$ be given.*

(a): $\mathcal{W}_\perp{}^\perp = \mathcal{W}$ *for every finite-dimensional subspace $\mathcal{W}$ of $\mathcal{V}^*$.*

(b): $\mathcal{B}_\perp{}^\perp = \mathrm{Lsp}\mathcal{B}$ *for every subset $\mathcal{B}$ of $\mathcal{V}^*$ such that $\mathrm{Lsp}\mathcal{B}$ is finite-dimensional; in particular, for every finite subset $\mathcal{B}$ of $\mathcal{V}^*$.*

Proof. 1. We first show that

(62.9) $\qquad B_\perp{}^\perp = \mathrm{Lsp}\mathcal{B} \quad$ for every finite subset $\mathcal{B}$ of $\mathcal{V}^*$.

Let a finite subset $\mathcal{B}$ of $\mathcal{V}^*$ be given. We consider the linear mapping $B \in \mathrm{Lin}(\mathcal{V}, \mathbb{F}^\mathcal{B})$ defined by the rule

$$(Bv)_\lambda := \lambda v \quad \text{for all } \lambda \in \mathcal{B} \text{ and } v \in \mathcal{V}.$$

(This B is the linear mapping identified with $\mathcal{B}$ self-indexed, by Proposition 31A; but we prefer not to burden the symbol $\mathcal{B}$ with an overload of meanings.) We find that

$$\mathrm{Null}(B|^{\mathrm{Rng}}) = \mathrm{Null}B = \{v \in \mathcal{V} \mid \forall \lambda \in \mathcal{B}, \lambda v = 0\} = \mathcal{B}_\perp.$$

Now let $\mu \in \mathcal{B}_\perp{}^\perp$ be given. Then $\mathrm{Null}\mu \supset \mathcal{B}_\perp = \mathrm{Null}(B|^{\mathrm{Rng}})$. Since $B|^{\mathrm{Rng}}$ is surjective, there is by Proposition 22B.R (or Theorem 22D) exactly one $\xi \in (\mathrm{Rng}B)^*$ such that $\mu = \xi(B|^{\mathrm{Rng}})$. By Corollary 21L (cf. Remark 52H), ξ is the restriction to $\mathrm{Rng}B$ of a suitable linear form on $\mathbb{F}^\mathcal{B}$. Since $\mathcal{B}$ is finite, we have $\mathbb{F}^{(\mathcal{B})} = \mathbb{F}^\mathcal{B}$, and by Proposition 61F this linear form is $a\cdot$ for a suitable $a \in \mathbb{F}^\mathcal{B}$. We conclude that

$$\mu = \xi(B|^{\mathrm{Rng}}) = (a\cdot|_{\mathrm{Rng}B})(B|^{\mathrm{Rng}}) = a\cdot B = \sum_{\lambda \in \mathcal{B}} a_\lambda \lambda \in \mathrm{Lsp}\mathcal{B}.$$

Since $\mu \in \mathcal{B}_\perp{}^\perp$ was arbitrary, we conclude that $\mathcal{B}_\perp{}^\perp \subset \mathrm{Lsp}\mathcal{B}$. By Corollary 62D,(a), we have $\mathrm{Lsp}\mathcal{B} \subset \mathcal{B}_\perp{}^\perp$, and therefore equality holds. This completes the proof of (62.9).

2. Let the finite-dimensional subspace $\mathcal{W}$ of $\mathcal{V}^*$ be given. Choose a finite subset $\mathcal{B}$ of $\mathcal{W}$ that spans $\mathcal{W}$. By (62.9) and Corollary 62D,(a) we then have $\mathcal{W}_\perp{}^\perp = (\mathrm{Lsp}\mathcal{B})_\perp{}^\perp = \mathcal{B}_\perp{}^\perp = \mathrm{Lsp}\mathcal{B} = \mathcal{W}$. This completes the proof of (a).

Finally, let $\mathcal{B}$ be a subset of $\mathcal{V}^*$ such that $\mathrm{Lsp}\mathcal{B}$ is finite-dimensional. By (a) and Corollary 62D,(a) we have $\mathcal{B}_\perp{}^\perp = (\mathrm{Lsp}\mathcal{B})_\perp{}^\perp = \mathrm{Lsp}\mathcal{B}$. This completes the proof of (b). ∎

62L. Remark. The finiteness assumptions in Theorem 62K may not be omitted. Indeed, let the set I be given, and consider the subset $\mathcal{B} := \{\delta_i^I \cdot \mid i \in I\}$ of $(\mathbb{F}^{(I)})^*$. It follows at once from (61.6) that $\mathcal{B}_\perp = \{0\}$, and therefore $\mathcal{B}_\perp{}^\perp = (\mathbb{F}^{(I)})^* = \mathbb{F}^I \cdot$ (cf. Remark 61G). However, $\mathrm{Lsp}\mathcal{B} = \mathbb{F}^{(I)} \cdot$, and so $\mathcal{B}_\perp{}^\perp = \mathrm{Lsp}\mathcal{B}$ if and only if $\mathbb{F}^{(I)} = \mathbb{F}^I$, hence if and only if I is finite and thus $\mathrm{Lsp}\mathcal{B}$ is finite-dimensional. ∎

•62M. Corollary. *Let the linear spaces $\mathcal{V}$ and $\mathcal{W}$ and the linear mapping $L \in \mathrm{Lin}(\mathcal{V}, \mathcal{W})$ be given. If $\mathrm{Rng}L^\top$ is finite-dimensional, then*

$$\mathrm{Rng}L^\top = (\mathrm{Null}L)^\perp.$$

Proof. By •Corollary 62I and Theorem 62K we have

$$\mathrm{Rng}L^\top = (\mathrm{Rng}L^\top)_\perp{}^\perp = (\mathrm{Null}L)^\perp. \quad ∎$$

The next two propositions deal with the relationships between annihilators and linear partitions.

62N. PROPOSITION. *Let the linear space $\mathcal{V}$ and the subspace $\mathcal{U}$ of $\mathcal{V}$ be given. Then there exists exactly one linear mapping $H \in \mathrm{Lin}((\mathcal{V}/\mathcal{U})^*, \mathcal{U}^\perp)$ such that $1_{\mathcal{U}^\perp \subset \mathcal{V}^*} H = \Omega_{\mathcal{V}/\mathcal{U}}{}^\top$; moreover, H is a linear isomorphism.*

Proof. $\Omega_{\mathcal{V}/\mathcal{U}}$ is surjective, hence $\Omega_{\mathcal{V}/\mathcal{U}}{}^\top$ is injective (Proposition 62F,(b)). We claim that $\mathrm{Rng}\Omega_{\mathcal{V}/\mathcal{U}}{}^\top = \mathcal{U}^\perp$; the conclusion then follows at once, with $H := (\Omega_{\mathcal{V}/\mathcal{U}}{}^\top)|^{\mathrm{Rng}}$.

Let $\lambda \in \mathcal{V}^*$ be given. Then $\lambda \in \mathrm{Rng}\Omega_{\mathcal{V}/\mathcal{U}}{}^\top$ if and only if $\lambda = \mu\Omega_{\mathcal{V}/\mathcal{U}}$ for a suitable $\mu \in (\mathcal{V}/\mathcal{U})^*$; since $\Omega_{\mathcal{V}/\mathcal{U}}$ is surjective, this will be the case if and only if $\mathrm{Null}\lambda \supset \mathrm{Null}\Omega_{\mathcal{V}/\mathcal{U}} = \mathcal{U}$ (Theorem 22D), i.e., if and only if $\lambda \in \mathcal{U}^\perp$. This establishes our claim and completes the proof. ∎

•**62O. PROPOSITION.** *Let the linear space $\mathcal{V}$ and the subspace $\mathcal{U}$ of $\mathcal{V}$ be given. Then there exists exactly one linear mapping $K \in \mathrm{Lin}(\mathcal{V}^*/\mathcal{U}^\perp, \mathcal{U}^*)$ such that $1_{\mathcal{U} \subset \mathcal{V}}{}^\top = K\Omega_{\mathcal{V}^*/\mathcal{U}^\perp}$; moreover, K is a linear isomorphism.*

Proof. By Proposition 62F,(a),

$$\mathrm{Null}(1_{\mathcal{U} \subset \mathcal{V}}{}^\perp) = (\mathrm{Rng}1_{\mathcal{U} \subset \mathcal{V}})^\perp = \mathcal{U}^\perp = \mathrm{Null}\Omega_{\mathcal{V}^*/\mathcal{U}^\perp}.$$

Moreover, $1_{\mathcal{U} \subset \mathcal{V}}{}^\top \in \mathrm{Lin}(\mathcal{V}^*, \mathcal{U}^*)$ is surjective (•Examples 61D,(b)). The conclusion follows by Corollary 22E. ∎

We turn to an examination of the relationships between annihilators and decompositions. In the remainder of this section, $\bigcap$ always stands for $\bigcap^{\mathcal{X}}$, where $\mathcal{X}$ is the obviously appropriate linear space – e.g., $\mathcal{X} := \mathcal{V}^*$ in the next proposition.

62P. PROPOSITION. *Let the linear space $\mathcal{V}$ and the finite decomposition $(\mathcal{U}_i \mid i \in I)$ of $\mathcal{V}$ be given, with family of idempotents $(E_i \mid i \in I)$. Then $(\bigcap_{i \in I \setminus \{j\}} \mathcal{U}_i^\perp \mid j \in I)$ is a decomposition of $\mathcal{V}^*$, with family of idempotents $(E_j{}^\top \mid j \in I)$.*

Proof. By Propositions 33E and 33F we have

$$(62.10) \qquad\qquad E_j E_i = \delta_{j,i}^I E_i \quad \text{for all } i, j \in I,$$

$$(62.11) \qquad\qquad \mathrm{Null}E_j = \sum_{i \in I \setminus \{j\}} \mathcal{U}_i \quad \text{for all } j \in I,$$

$$(62.12) \qquad\qquad \sum_{i \in I} E_i = 1_{\mathcal{V}}.$$

For each $j \in I$, $1_{\mathcal{V}} - E_j$ is also idempotent, so that its range is the set of its fixed points; thus

$$(62.13) \qquad \mathrm{Null}E_j = \mathrm{Null}(1_{\mathcal{V}} - (1_{\mathcal{V}} - E_j)) = \mathrm{Rng}(1_{\mathcal{V}} - E_j) \quad \text{for all } j \in I.$$

Now $(E_j{}^\top \mid j \in I)$ is a family in $\mathrm{Lin}\mathcal{V}^*$. By (62.10), (62.12), and Proposition 61C we have

$$(62.14) \qquad E_j{}^\top E_i{}^\top = (E_i E_j)^\top = \delta_{i,j}^I E_j{}^\top = \delta_{i,j}^I E_i{}^\top \quad \text{for all } i, j \in I,$$

$$(62.15) \qquad \sum_{j\in I} E_j{}^\top = (\sum_{j\in I} E_j)^\top = 1_\mathcal{V}{}^\top = 1_{\mathcal{V}^*}.$$

For each $j \in I$, $E_j{}^\top$ is idempotent, by (62.14), and hence its range is the set of its fixed points. Using Propositions 61C and 62F and Corollary 62E,(b), we find, from (62.11) and (62.13),

$$\mathrm{Rng}E_j{}^\top = \mathrm{Null}(1_{\mathcal{V}^*} - E_j{}^\top) = \mathrm{Null}(1_\mathcal{V} - E_j)^\top = (\mathrm{Rng}(1_\mathcal{V} - E_j))^\perp =$$
$$= (\mathrm{Null}E_j)^\perp = (\sum_{i\in I\setminus\{j\}} \mathcal{U}_i)^\perp = \bigcap_{i\in I\setminus\{j\}} \mathcal{U}_i^\perp \quad \text{for all } j \in I.$$

From (12.14), (12.15), (12.16) it follows by Proposition 33F that $(\bigcap_{i\in I\setminus\{j\}} \mathcal{U}_i^\perp \mid j \in I)$ is a decomposition of $\mathcal{V}^*$ and that $(E_j{}^\perp \mid j \in I)$ is the family of idempotents of this decomposition. ∎

62Q. COROLLARY. *Let the linear space $\mathcal{V}$ and the subspaces $\mathcal{U}$, $\mathcal{U}'$ of $\mathcal{V}$ be given. If $\mathcal{U}$ and $\mathcal{U}'$ are supplementary in $\mathcal{V}$, then $\mathcal{U}^\perp$ and $\mathcal{U}'^\perp$ are supplementary in $\mathcal{V}^*$.*

62R. REMARK. The assumption that I is finite may not be omitted in Proposition 62P. (It may, however, be replaced by the assumption that the support of $(\mathcal{U}_i \mid i \in I)$ is finite; we leave this adjustment to the reader.) Let the infinite set I be given, and consider the decomposition $(\mathbb{F}\delta_i^I \mid i \in I)$ of the linear space $\mathbb{F}^{(I)}$. By Corollary 62D we have

$$(62.17) \qquad \bigcap_{i\in I\setminus\{j\}} (\mathbb{F}\delta_i^I)^\perp = \bigcap_{i\in I\setminus\{j\}} \{\delta_i^I\}^\perp = \{\delta_i^I \mid i \in I\setminus\{j\}\}^\perp \quad \text{for all } j \in I.$$

It follows that, for every $j \in I$ and $a \in \mathbb{F}^I$,

$$a\boldsymbol{\cdot} \in \{\delta_i^I \mid i \in I\setminus\{j\}\}^\perp \;\Leftrightarrow\; (\forall i \in I\setminus\{j\},\, a\boldsymbol{\cdot}\delta_i^I = 0) \;\Leftrightarrow$$
$$\Leftrightarrow\; (\forall i \in I\setminus\{j\},\, a_i = 0) \quad \Leftrightarrow\; a \in \mathbb{F}\delta_j^I \;\Leftrightarrow\; a\boldsymbol{\cdot} \in \mathbb{F}\delta_j^I\boldsymbol{\cdot} \; .$$

Using Proposition 61F and combining the preceding equivalence with (62.17), we conclude that

$$\bigcap_{i\in I\setminus\{j\}} (\mathbb{F}\delta_i^I)^\perp = \mathbb{F}\delta_j^I \quad \text{for all } j \in I.$$

If the family of these subspaces of $(\mathbb{F}^{(I)})^*$ were a decomposition of $(\mathbb{F}^{(I)})^*$, then $(\mathbb{F}\delta_j^I \mid j \in I)$ would be a decomposition of $\mathbb{F}^I$, and then $\mathbb{F}^I = \sum_{i \in I} \mathbb{F}\delta_j^I = \mathbb{F}^{(I)}$; but I was infinite, so this equality does not hold. ∎

•**62S. PROPOSITION.** *Let the linear space $\mathcal{V}$ and the subspaces $\mathcal{U}$, $\mathcal{U}'$ of $\mathcal{V}$ be given. Then $\mathcal{U}$ and $\mathcal{U}'$ are supplementary in $\mathcal{V}$ if and only if $\mathcal{U}^\perp$ and $\mathcal{U}'^\perp$ are supplementary in $\mathcal{V}^*$.*

Proof. The "only if" part follows from Corollary 62Q. To prove the "•if" part, assume that $\mathcal{U}^\perp$ and $\mathcal{U}'^\perp$ are supplementary in $\mathcal{V}^*$. By •Theorem 62H,(a) and Corollary 62E, (b) we have

$$\mathcal{U} \cap \mathcal{U}' = \mathcal{U}^\perp{}_\perp \cap \mathcal{U}'^\perp{}_\perp = (\mathcal{U}^\perp + \mathcal{U}'^\perp)_\perp = \mathcal{V}^*_\perp = \{0\}$$
$$\mathcal{U} + \mathcal{U}' = (\mathcal{U} + \mathcal{U}')^\perp{}_\perp = (\mathcal{U}^\perp \cap \mathcal{U}'^\perp)_\perp = \{0\}_\perp = \mathcal{V}. \blacksquare$$

▼ We conclude this section by obtaining, in •Proposition 62U, a converse of Proposition 62P, without any finiteness assumption. (The "if" part of •Proposition 62S is a special case of this converse.)

62T. LEMMA. *Let the linear space $\mathcal{X}$ and the family $(\mathcal{Y}_i \mid i \in I)$ of subspaces of $\mathcal{X}$ be given. Assume that I is neither empty nor a singleton, and that the family $(\bigcap_{i \in I\setminus\{j\}} \mathcal{Y}_i \mid i \in I)$ is a decomposition of $\mathcal{X}$. Then*

$$(62.18) \qquad \mathcal{Y}_k = \sum_{j \in I\setminus\{k\}} \bigcap_{i \in I\setminus\{j\}} \mathcal{Y}_i \quad \text{for all } k \in I,$$

and therefore $\mathcal{Y}_k$ is a supplement of $\bigcap_{i \in I\setminus\{k\}} \mathcal{Y}_i$ in $\mathcal{X}$ for every $k \in I$.

Proof. Set $\mathcal{Z}_j := \bigcap_{i \in I\setminus\{j\}} \mathcal{Y}_i$ for every $j \in I$. Thus $(\mathcal{Z}_j \mid j \in I)$ is a decomposition of $\mathcal{X}$. Since I is neither empty nor a singleton, we may choose $p, q \in I$ such that $p \neq q$, and find

$$(62.19) \qquad \bigcap_{i \in I} \mathcal{Y}_i = \mathcal{Z}_p \cap \mathcal{Z}_q = \{0\}.$$

Let $k \in I$ be given. For every $j \in I\setminus\{k\}$ we have $k \in I\setminus\{j\}$, and therefore

$$\mathcal{Z}_j \subset \mathcal{Y}_k \quad \text{for all } j \in I\setminus\{k\}.$$

It follows at once that

$$(62.20) \qquad \mathcal{Y}_k \supset \sum_{j \in I\setminus\{k\}} \mathcal{Z}_j.$$

Let $(E_j \mid j \in I)$ be the family of idempotents of the decomposition $(\mathcal{Z}_j \mid j \in I)$ of $\mathcal{X}$. Let $x \in \mathcal{Y}_k$ be given. Then $(E_j x \mid j \in I)$ has finite support, and $\sum_{j \in I} E_j x = x$ (Proposition 33F). It follows by (62.20) that

$$E_k x = x - \sum_{j \in I \setminus \{k\}} E_j x \in \mathcal{Y}_k.$$

But $E_k x \in \mathcal{Z}_k$; therefore, by (62.19),

$$E_k x \in \mathcal{Y}_k \cap \mathcal{Z}_k = \bigcap_{i \in I} \mathcal{Y}_i = \{0\},$$

and consequently $x = \sum_{j \in I \setminus \{k\}} E_j x \in \sum_{j \in I \setminus \{k\}} \mathcal{Z}_j$. Since $x \in \mathcal{Y}_k$ was arbitrary, we conclude that

$$\mathcal{Y}_k \subset \sum_{j \in I \setminus \{k\}} \mathcal{Z}_j.$$

Combination of this with (62.20) yields (62.18), since $k \in I$ was arbitrary.

The final clause of the conclusion then follows by Theorem 33A. ■

•**62U. PROPOSITION.** *Let the linear space $\mathcal{V}$ and the family $(\mathcal{U}_i \mid i \in I)$ of subspaces of $\mathcal{V}$ be given, and assume that I is not a singleton. If the family $(\bigcap_{i \in I \setminus \{j\}} \mathcal{U}_i^\perp \mid j \in I)$ is a decomposition of $\mathcal{V}^*$, then $(\mathcal{U}_i \mid i \in I)$ is a decomposition of $\mathcal{V}$.*

Proof. We set $\mathcal{W}_j := \bigcap_{i \in I \setminus \{j\}} \mathcal{U}_i^\perp$ for every $j \in I$, and assume that $(\mathcal{W}_j \mid j \in I)$ is a decomposition of $\mathcal{V}^*$.

If I is empty, so is this decomposition, and hence $\mathcal{V}^* = \{0\}$. But then we have, by Proposition 62B,(b) and •Theorem 62H,(a), $\mathcal{V} = \{0\}_\perp = \mathcal{V}^*{}_\perp = \{0\}$; and the empty family $(\mathcal{U}_i \mid i \in I)$ is indeed a decomposition of $\mathcal{V}$.

We may therefore assume from now on that I is neither empty nor a singleton. It follows by Lemma 62T that $\mathcal{U}_j{}^\perp$ is a supplement of $\mathcal{W}_j$ in $\mathcal{V}^*$ for every $j \in I$. For every $j \in I$, $(\sum_{i \in I \setminus \{j\}} \mathcal{U}_i)^\perp = \mathcal{W}_j$, by Corollary 62E,(b); hence by •Proposition 62R, $\mathcal{U}_j$ is a supplement of $\sum_{i \in I \setminus \{j\}} \mathcal{U}_i$ in $\mathcal{V}$ for every $j \in I$. We conclude, by Theorem 33A, that $(\mathcal{U}_i \mid i \in I)$ is a decomposition of $\mathcal{V}$. ■

62V. REMARK. The assertion of •Proposition 62U does not remain valid in general if the assumption that I is not a singleton is omitted. Indeed, suppose that $I := \{k\}$, that $\mathcal{V} \neq \{0\}$ (e.g., $\mathcal{V} := \mathbb{F}$), and that $\mathcal{U}_k := \{0\}$. Then $(\mathcal{U}_i \mid i \in I)$ is not a decomposition of $\mathcal{V}$; but $\mathcal{W}_k := \bigcap_{i \in I \setminus \{k\}} \mathcal{U}_i^\perp = \bigcap_{i \in \varnothing} \mathcal{U}_i^\perp = \mathcal{V}^*$, so that $(\mathcal{W}_j \mid j \in I)$ is a decomposition of $\mathcal{V}^*$. ■

63. Higher-order dual spaces and transposes

Given a linear space $\mathcal{V}$, we have not only its dual space $\mathcal{V}^*$, but also the dual space $\mathcal{V}^{**}$ of $\mathcal{V}^*$, called the **second dual space of** $\mathcal{V}$ and so on. (We shall not need the easily supplied formal recursive definition of the "nth dual space of $\mathcal{V}$".) We describe, for each linear space $\mathcal{V}$, an important linear mapping from $\mathcal{V}$ to $\mathcal{V}^{**}$.

Let the linear space $\mathcal{V}$ be given. For each $u \in \mathcal{V}$, the evaluation mapping $\mathrm{ev}^{\mathrm{Map}(\mathcal{V},\mathbb{F})}{}_u : \mathrm{Map}(\mathcal{V},\mathbb{F}) \to \mathbb{F}$ is linear (Examples 12G,(b)), and hence so is its restriction $\mathrm{ev}^{\mathcal{V}^*}{}_u : \mathcal{V}^* \to \mathbb{F}$ to the subspace $\mathcal{V}^*$ of $\mathrm{Map}(\mathcal{V},\mathbb{F})$; this restriction is thus a member of $\mathcal{V}^{**}$. We may therefore define the adjustment $\mathrm{Ev}_{\mathcal{V}} := \mathrm{ev}^{\mathcal{V}^*}|^{\mathcal{V}^{**}} : \mathcal{V} \to \mathcal{V}^{**}$ of the evaluation family $\mathrm{ev}^{\mathcal{V}^*}$; this mapping satisfies the following formula, and is indeed characterized by it:

(63.1) $(\mathrm{Ev}_{\mathcal{V}}(u))\lambda = \lambda u$ for all $u \in \mathcal{V}$ and $\lambda \in \mathcal{V}^*$.

63A. Proposition. *For every linear space $\mathcal{V}$, the mapping $\mathrm{Ev}_{\mathcal{V}} : \mathcal{V} \to \mathcal{V}^{**}$ is linear.*

63B. Remark. There is another approach to the linear mapping $\mathrm{Ev}_{\mathcal{V}}$. Let the linear space $\mathcal{V}$ be given. Let $u \in \mathcal{V}$ be given; then $(u\otimes)^{\mathsf{T}} \in \mathrm{Lin}(\mathcal{V}^*,\mathbb{F}^*)$. Under the identification of $\mathrm{Lin}(\mathcal{V}^*,\mathbb{F}^*)$ with $\mathrm{Lin}(\mathcal{V}^*,\mathbb{F}) = \mathcal{V}^{**}$ (Remark 61B), we have

$$(u\otimes)^{\mathsf{T}}\lambda = \lambda u\otimes = \lambda u = (\mathrm{Ev}_{\mathcal{V}}u)\lambda \quad \text{for all } u \in \mathcal{V}.$$

Thus

(63.2) $(u\otimes)^{\mathsf{T}} = \mathrm{Ev}_{\mathcal{V}}u$ for all $u \in \mathcal{V}$.

With this observation, the linearity of $\mathrm{Ev}_{\mathcal{V}}$ becomes evident. ∎

Let the linear spaces $\mathcal{V}$ and $\mathcal{W}$ be given. For every linear mapping $L \in \mathrm{Lin}(\mathcal{V},\mathcal{W})$ we have $L^{\mathsf{TT}} = (L^{\mathsf{T}})^{\mathsf{T}} \in \mathrm{Lin}(\mathcal{V}^{**},\mathcal{W}^{**})$, and this linear mapping may be called the **second transpose of** L.

63C. Proposition. *Let the linear spaces $\mathcal{V}$ and $\mathcal{W}$ be given. Then*

$$L^{\mathsf{TT}}\mathrm{Ev}_{\mathcal{V}} = \mathrm{Ev}_{\mathcal{W}}L \quad \text{for all } L \in \mathrm{Lin}(\mathcal{V},\mathcal{W}).$$

Proof. Let $L \in \mathrm{Lin}(\mathcal{V},\mathcal{W})$ be given. Using (63.1) we have

$$(L^{\mathsf{TT}}\mathrm{Ev}_{\mathcal{V}}u)\mu = (\mathrm{Ev}_{\mathcal{V}}u)(L^{\mathsf{T}}\mu) = (L^{\mathsf{T}}\mu)u = \mu Lu = (\mathrm{Ev}_{\mathcal{W}}Lu)\mu$$
$$\text{for all } \mu \in \mathcal{W}^* \text{ and } u \in \mathcal{V}.$$

Therefore $L^{\mathsf{TT}}\mathrm{Ev}_{\mathcal{V}}u = \mathrm{Ev}_{\mathcal{W}}Lu$ for all $u \in \mathcal{V}$, and the assertion follows. ∎

The conclusion of Proposition 63C may be expressed by the commutativity of the following diagram:

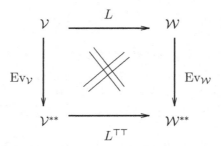

63D. PROPOSITION. *For every linear space* $\mathcal{V}$ *we have* $\mathrm{Ev}_{\mathcal{V}}{}^{\mathsf{T}}\mathrm{Ev}_{\mathcal{V}^*} = 1_{\mathcal{V}^*}$. *In particular,* $\mathrm{Ev}_{\mathcal{V}^*}$ *is injective and* $\mathrm{Ev}_{\mathcal{V}}{}^{\mathsf{T}}$ *is surjective.*

Proof. Let the linear space $\mathcal{V}$ be given. We have

$$(\mathrm{Ev}_{\mathcal{V}}{}^{\mathsf{T}}\mathrm{Ev}_{\mathcal{V}^*}\lambda)v = (\mathrm{Ev}_{\mathcal{V}^*}\lambda)(\mathrm{Ev}_{\mathcal{V}}v) = (\mathrm{Ev}_{\mathcal{V}}v)\lambda = \lambda v \quad \text{for all } v \in \mathcal{V} \text{ and } \lambda \in \mathcal{V}^*.$$

Therefore $\mathrm{Ev}_{\mathcal{V}}{}^{\mathsf{T}}\mathrm{Ev}_{\mathcal{V}^*}\lambda = \lambda$ for all $\lambda \in \mathcal{V}^*$, and the assertion follows. ∎

We just saw that $\mathrm{Ev}_{\mathcal{V}^*}$ is injective. Actually, $\mathrm{Ev}_{\mathcal{V}}$ is injective for *every* linear space $\mathcal{V}$, as we now show. In view of this fact, $\mathrm{Ev}_{\mathcal{V}}$ is often called the **canonical injection of** $\mathcal{V}$.

•**63E. COROLLARY.** *For every linear space* $\mathcal{V}$, $\mathrm{Ev}_{\mathcal{V}}$ *is injective.*

Proof. By Proposition 63D, $\mathrm{Ev}_{\mathcal{V}}{}^{\mathsf{T}}$ is surjective; by •Corollary 62I, $\mathrm{Ev}_{\mathcal{V}}$ is injective. ∎

63F. COROLLARY. *Let the linear spaces* $\mathcal{V}$ *and* $\mathcal{W}$ *be given. Then*

$$\mathrm{Ev}_{\mathcal{W}}{}^{\mathsf{T}}L^{\mathsf{TT}}\mathrm{Ev}_{\mathcal{V}} = L \quad \text{for all } L \in \mathrm{Lin}(\mathcal{V}, \mathcal{W}^*).$$

Proof. By Propositions 63C and 63D we have

$$\mathrm{Ev}_{\mathcal{W}}{}^{\mathsf{T}}L^{\mathsf{TT}}\mathrm{Ev}_{\mathcal{V}} = \mathrm{Ev}_{\mathcal{W}}{}^{\mathsf{T}}\mathrm{Ev}_{\mathcal{W}^*}L = 1_{\mathcal{W}^*}L = L \quad \text{for every } L \in \mathrm{Lin}(\mathcal{V}, \mathcal{W}^*). ∎$$

The next results describe the relationship between $\mathrm{Ev}_{\mathcal{V}}$ and annihilators and pre-annihilators.

63G. PROPOSITION. *Let the linear space* $\mathcal{V}$ *be given. Then*

$$\mathrm{Ev}_{\mathcal{V}}{}^{<}(\mathcal{B}^{\perp}) = \mathcal{B}_{\perp} \quad \text{for every subset } \mathcal{B} \text{ of } \mathcal{V}^*.$$

Proof. Let the subset $\mathcal{B}$ of $\mathcal{V}^*$ be given. For every $v \in \mathcal{V}$ we have the chain of equivalences

$$v \in \mathrm{Ev}_{\mathcal{V}}{}^{<}(\mathcal{B}^{\perp}) \quad \Leftrightarrow \quad \mathrm{Ev}_{\mathcal{V}}v \in \mathcal{B}^{\perp} \quad \Leftrightarrow \quad (\forall \lambda \in \mathcal{B}, \; (\mathrm{Ev}_{\mathcal{V}}v)\lambda = 0) \quad \Leftrightarrow$$

$$\Leftrightarrow \quad (\forall \lambda \in \mathcal{B}, \; \lambda v = 0) \quad \Leftrightarrow v \in \mathcal{B}_{\perp}. ∎$$

•**63H. COROLLARY.** *Let the linear space $\mathcal{V}$ be given. Then*

$$\text{Ev}_{\mathcal{V}}{}^{<}(\mathcal{A}^{\perp\perp}) = \text{Lsp}\mathcal{A} \quad \textit{for every subset } \mathcal{A} \textit{ of } \mathcal{V}.$$

Proof. Apply Proposition 63G to $\mathcal{B} := \mathcal{A}^{\perp}$ and use •Theorem 62H,(b) to find $\text{Ev}_{\mathcal{V}}(\mathcal{A}^{\perp\perp}) = \mathcal{A}^{\perp}{}_{\perp} = \text{Lsp}\mathcal{A}$ for all subsets $\mathcal{A}$ of $\mathcal{V}$. ∎

We derive some interesting consequences from Propositions 63C and 63D and Corollary 63F.

63I. PROPOSITION. *Let the linear space $\mathcal{V}$ be given. Then $\text{Ev}_{\mathcal{V}*}\text{Ev}_{\mathcal{V}}{}^{\top} \in \text{Lin}\mathcal{V}^{***}$ is idempotent; the null-space and the range of this linear mapping are $(\text{Rng Ev}_{\mathcal{V}})^{\perp}$ and $\text{Rng Ev}_{\mathcal{V}*}$, respectively, and these subspaces of $\mathcal{V}^{***}$ are supplementary in $\mathcal{V}^{***}$.*

Proof. It follows at once from Proposition 63D that $\text{Ev}_{\mathcal{V}*}\text{Ev}_{\mathcal{V}}{}^{\top}$ is idempotent. Since $\text{Ev}_{\mathcal{V}*}$ is injective, we have, by Proposition 62F,(a), $\text{Null}(\text{Ev}_{\mathcal{V}*}\text{Ev}_{\mathcal{V}}{}^{\top}) =$ $= \text{Null Ev}_{\mathcal{V}}{}^{\top} = (\text{Rng Ev}_{\mathcal{V}})^{\perp}$. Since $\text{Ev}_{\mathcal{V}}{}^{\top}$ is surjective, we have $\text{Rng }(\text{Ev}_{\mathcal{V}*}\text{Ev}_{\mathcal{V}}{}^{\top}) =$ $= \text{Rng Ev}_{\mathcal{V}*}$. These subspaces are supplementary in $\mathcal{V}^{***}$ by Proposition 15C. ∎

63J. THEOREM. *Let the family of linear spaces $(\mathcal{V}_i \mid i \in I)$ be given. Let $\mathcal{S}$ be the coproduct space and $(s_i \mid i \in I)$ the family of insertions of a linear coproduct of $(\mathcal{V}_i \mid i \in I)$. Then $\mathcal{S}^*$ is the product space and $(s_i{}^{\top} \mid i \in I)$ the family of projections of a linear product of the family of linear spaces $(\mathcal{V}_i^* \mid i \in I)$.*

Proof. Let the linear space $\mathcal{W}$ and the family of linear mappings $(L_i \mid i \in I) \in \underset{i \in I}{\bigtimes} \text{Lin}(\mathcal{W}, \mathcal{V}_i{}^*)$ be given. By the coproduct assumptions, there is exactly one $K \in \text{Lin}(\mathcal{S}, \mathcal{W}^*)$ such that

$$(63.3) \qquad\qquad Ks_i = L_i{}^{\top}\text{Ev}_{\mathcal{V}_i} \quad \text{for all } i \in I.$$

We are to show that there is exactly one $L \in \text{Lin}(\mathcal{W}, \mathcal{S}^*)$ such that

$$(63.4) \qquad\qquad s_i{}^{\top}L = L_i \quad \text{for all } i \in I.$$

Let $L \in \text{Lin}(\mathcal{W}, \mathcal{S}^*)$ be given, and assume that L satisfies (63.4). We then have, as indicated in the commutative diagram (for each $i \in I$)

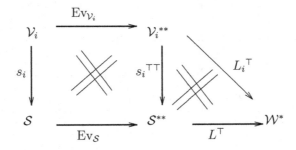

the following consequences of (63.4) and Proposition 63C:

$$L^{\mathsf{T}}\mathrm{Ev}_{\mathcal{S}}s_i = L^{\mathsf{T}}s_i{}^{\mathsf{TT}}\mathrm{Ev}_{\mathcal{V}_i} = (s_i{}^{\mathsf{T}}L)^{\mathsf{T}}\mathrm{Ev}_{\mathcal{V}_i} = L_i{}^{\mathsf{T}}\mathrm{Ev}_{\mathcal{V}_i} \quad \text{for all } i \in I.$$

Comparison with (63.3) shows that we must have $L^{\mathsf{T}}\mathrm{Ev}_{\mathcal{S}} = K$. By Corollary 63F we must then have

$$L = \mathrm{Ev}_{\mathcal{S}}{}^{\mathsf{T}}L^{\mathsf{TT}}\mathrm{Ev}_{\mathcal{W}} = (L^{\mathsf{T}}\mathrm{Ev}_{\mathcal{S}})^{\mathsf{T}}\mathrm{Ev}_{\mathcal{W}} = K^{\mathsf{T}}\mathrm{Ev}_{\mathcal{W}}.$$

This shows that there is at most one $L \in \mathrm{Lin}(\mathcal{W}, \mathcal{S}^*)$ satisfying (63.4).

It remains to show that $L := K^{\mathsf{T}}\mathrm{Ev}_{\mathcal{W}}$ does indeed satisfy (63.4). This is done using (63.3) and Corollary 63F once more:

$$s_i{}^{\mathsf{T}}L = s_i{}^{\mathsf{T}}K^{\mathsf{T}}\mathrm{Ev}_{\mathcal{W}} = (Ks_i)^{\mathsf{T}}\mathrm{Ev}_{\mathcal{W}} = (L_i{}^{\mathsf{T}}\mathrm{Ev}_{\mathcal{V}_i})^{\mathsf{T}}\mathrm{Ev}_{\mathcal{W}} = \mathrm{Ev}_{\mathcal{V}_i}{}^{\mathsf{T}}L_i{}^{\mathsf{TT}}\mathrm{Ev}_{\mathcal{W}} = L_i$$
$$\text{for all } i \in I,$$

as desired. ∎

Theorem 63J contains special cases closely related to Propositions 61F and 62P. The detection of these relationships is left to the reader.

64. Biorthogonal families

Let the linear space V be given. The families $a \in V^I$ and $\alpha \in V^{*I}$ (with the same index set) are said to be **biorthogonal** if

$$\alpha_i a_j = \delta^I_{i,j} \quad \text{for all } i, j \in I;$$

in that case we also say that the pair (a, α) is **biorthogonal**, and that α is **biorthogonal to** a.

64A. PROPOSITION. *Let the linear space V and families $a \in V^I$ and $\alpha \in V^{*I}$ be given. If a and α are biorthogonal, then each of these families is linearly independent.*

Proof. Assume that a and α are biorthogonal. Let $s \in \text{Null lc}^V_a$ be given. Then

$$0 = \alpha_j 0 = \alpha_j \text{lc}_a s = \sum_{i \in I} s_i \alpha_j a_i = \sum_{i \in I} s_i \delta^I_{j,i} = s_j \quad \text{for all } j \in I,$$

so that $s = 0$. Thus Null $\text{lc}^V_a = \{0\}$, and hence a is linearly independent. The proof of the linear independence of α is entirely similar. ∎

64B. THEOREM. *Let the linear space V and a basis $b \in V^I$ of V be given. There exists exactly one family $b^* \in V^{*I}$ such that b^* is biorthogonal to b.*

Proof. This is an immediate consequence of Theorem 43F,(c). ∎

Whenever $b \in V^I$ is known to be a basis of the linear space V, b^* shall denote the unique family in V^* that is biorthogonal to b.

64C. COROLLARY. *Let the linear space V and a basis $b \in V^I$ be given. Then* $\text{Supp} b^* v$ *is finite for every $v \in V$, and*

$$(64.1) \qquad v = \sum_{i \in I} (b^*_i v) b_i = \sum_{i \in I} (b_i \otimes b^*_i) v \quad \text{for every } v \in V.$$

Proof. Let $v \in V$ be given and set $s := (\text{lc}^V_b)^{-1} v \in \mathbb{F}^{(I)}$, so that

$$(64.2) \qquad v = \text{lc}_b s = \sum_{i \in I} s_i b_i.$$

Then

$$(64.3) \qquad b^*_j v = \sum_{i \in I} s_i b^*_j b_i = \sum_{i \in I} s_i \delta^I_{j,i} = s_j \quad \text{for all } j \in I.$$

Thus $b^* v = s \in \mathbb{F}^{(I)}$ and (64.1) follows upon substitution of (64.3) into (64.2). ∎

64D. REMARKS. (a): Corollary 64C shows that, for a given linear space $\mathcal{V}$ and basis $b \in \mathcal{V}^I$ of $\mathcal{V}$, the linear mapping $b^* : \mathcal{V} \to \mathbb{F}^{(I)}$ (cf. Proposition 31A) is precisely $(\mathrm{lc}_b^{\mathcal{V}})^{-1}$.

(b): For every set I, the unique family in $(\mathbb{F}^{(I)})^*$ that is biorthogonal to the basis $\delta^I = (\delta_i^I \mid i \in I)$ of $\mathbb{F}^{(I)}$ is the family $(\delta_i^I \cdot \mid i \in I)$. ∎

64E. COROLLARY. *Let the linear spaces $\mathcal{V}$ and $\mathcal{W}$, the basis $b \in \mathcal{V}^I$ of $\mathcal{V}$, the family $y \in \mathcal{W}^I$, and the linear mapping $L \in \mathrm{Lin}(\mathcal{V}, \mathcal{W})$ be given. Then $Lb = y$ if and only if*

$$(64.4) \qquad Lv = \sum_{i \in I} (y_i \otimes b_i^*)v \quad \text{for all } v \in \mathcal{V}.$$

Proof. If (64.4) holds, then

$$Lb_j = \sum_{i \in I} (b_i^* b_j) y_i = \sum_{i \in I} \delta_{i,j}^I y_i = y_j \quad \text{for all } j \in I,$$

so that $Lb = y$. If, conversely, $Lb = y$, we have, by Corollary 64C,

$$Lv = \sum_{i \in I} (Lb_i \otimes b_i^*)v = \sum_{i \in I} (y_i \otimes b_i^*)v \quad \text{for all } v \in \mathcal{V}. \blacksquare$$

64F. PROPOSITION. *Let the linear space $\mathcal{V}$ and the basis $b \in \mathcal{V}^I$ of $\mathcal{V}$ be given. Then b^* is a basis of $\mathcal{V}^*$ if and only if I is finite, i.e., if and only if $\mathcal{V}$ is finite-dimensional.*

Proof. By Remark 64D,(a) we have $b^* = (\mathrm{lc}_b^{\mathcal{V}})^{-1} \in \mathrm{Lis}(\mathcal{V}, \mathbb{F}^{(I)})$. Therefore $(b^*)^\top \in \mathrm{Lis}((\mathbb{F}^{(I)})^*, \mathcal{V}^*)$, by Proposition 61C,(d). We find that

$$((b^*)^\top (\delta_i^I \cdot)) b_j = \delta_i^I \cdot (b^* b_j) = \delta_i^I \cdot \delta_j^I = \delta_{i,j}^I \quad \text{for all } i, j \in I,$$

and therefore $(b^*)^\top (\delta_i^I \cdot) = b_i^*$ for all $i \in I$ (Theorem 64B). We conclude, by Proposition 43D, that b^* is a basis of $\mathcal{V}^*$ if and only if $(\delta_i^I \cdot \mid i \in I)$ is a basis of $(\mathbb{F}^{(I)})^*$, and hence if and only if $\delta^I = (\delta_i^I \mid i \in I)$ is a basis of $\mathbb{F}^I$ (Proposition 61F). But δ^I is a basis of the subspace $\mathbb{F}^{(I)}$ of $\mathbb{F}^I$; therefore b^* is a basis of $\mathcal{V}^*$ if and only if $\mathbb{F}^{(I)} = \mathbb{F}^I$, i.e., if and only if I is finite. By Corollary 53C, (i)⇒(v), this condition is equivalent to $\mathcal{V}$ being finite-dimensional. ∎

Given a linear space $\mathcal{V}$ and a family $a \in \mathcal{V}^I$ [a family $\alpha \in \mathcal{V}^{*I}$], when does there exist a family $\alpha \in \mathcal{V}^{*I}$ [a family $a \in \mathcal{V}^I$] such that a and α are biorthogonal? By Proposition 64A, a necessary condition is that a [that α] be linearly independent. The next results show that this condition is [sometimes] also sufficient.

•**64G. PROPOSITION.** *Let the linear space $\mathcal{V}$ and the family $a \in \mathcal{V}^I$ be given. Then a family $\alpha \in \mathcal{V}^{*I}$ that is biorthogonal to a exists •if and only if a is linearly independent.*

Proof. The "only if" part follows by Proposition 64A. The "•if" part follows by •Corollary 43H. ∎

64H. Proposition. *Let the linear space $\mathcal{V}$ and the finite family $\alpha \in \mathcal{V}^{*I}$ be given. Then a family $a \in \mathcal{V}^I$ such that α is biorthogonal to a exists if and only if α is linearly independent.*

Proof. The "only if" part follows by Proposition 64A. To prove the "if" part, assume that α is linearly independent. We shall prove by special induction that $P(J)$ holds for all $J \in \mathfrak{F}(I) = \mathfrak{P}(I)$, where

$$P(J) :\Leftrightarrow \text{(There exists a family } a \in \mathcal{V}^J \text{ such that } a|_J \text{ is biorthogonal to } \alpha).$$

In particular, this will establish $P(I)$, which is the desired conclusion.

$P(\emptyset)$ holds trivially. Let $J \in \mathfrak{P}^\times(I)$ and $j \in J$ be given, and assume that $P(J\backslash\{j\})$ holds. We may therefore choose $a' \in \mathcal{V}^{J\backslash\{j\}}$ such that $\alpha|_{J\backslash\{j\}}$ is biorthogonal to a'.

Since α is linearly independent, so is $\alpha|_J$. By Corollary 42I, and by Theorem 62K,(b), we have

$$\alpha_j \notin \mathrm{Lsp}\ \mathrm{Rng}(\alpha|_{J\backslash\{j\}}) = (\mathrm{Rng}(\alpha|_{J\backslash\{j\}}))\perp^\perp.$$

We may therefore choose $u \in (\mathrm{Rng}(\alpha|_{J\backslash\{j\}}))\perp$ such that $\alpha_j u \neq 0$; replacing u by $(1/(\alpha_j u))u$ if necessary, we may assume without loss that $\alpha_j u = 1$.

We now define $a \in \mathcal{V}^J$ by

$$a_i := \begin{cases} u & \text{if } i = j \\ a_i' - (\alpha_j a_i')u & \text{if } i \in J\backslash\{j\}. \end{cases}$$

Direct verification – using the fact that $a_i u = 0$ for all $i \in J\backslash\{j\}$ – then shows that $\alpha|_J$ is biorthogonal to a. This completes the induction step. ∎

64I. Remark. The assumption that the family α is finite may not be omitted in Proposition 64H. Indeed, let the infinite set I be given, and choose an object ω such that $\omega \notin I$. Consider the family $\alpha \in ((\mathbb{F}^{(I)})^*)^{I\cup\{\omega\}}$ defined by

$$\alpha_i := \begin{cases} \delta_i^I \cdot & \text{if } i \in I \\ 1\cdot & \text{if } i = \omega \end{cases}$$

(where $1\cdot$ is the constant family with only term 1). Since I is infinite, α is linearly independent. However, it follows from (61.6) that if $a \in (\mathbb{F}^{(I)})^{I\cup\{\omega\}}$ is such that α is biorthogonal to a we must have $a_i = \delta_i^I$ for all $i \in I$; but then $\alpha_\omega a_i = 1\cdot\delta_i^I = 1 \neq 0$ for all $i \in I$, and α is not biorthogonal to a after all. Thus there is *no* $a \in (\mathbb{F}^{(I)})^{I\cup\{\omega\}}$ to which α is biorthogonal. ∎

Chapter 7

DUALITY AND FINITE DIMENSION

71. Duality for finite-dimensional spaces

The fundamental result concerning duality for finite-dimensional linear spaces is the fact that the dual space of such a space is also finite-dimensional and has the same dimension.

71A. THEOREM. *Let the finite-dimensional linear space $\mathcal{V}$ be given. Then $\mathcal{V}^*$ is finite-dimensional, and* $\dim \mathcal{V}^* = \dim \mathcal{V}$.

Proof. It follows from Corollary 53O that $\mathcal{V}^*$ is finite-dimensional, with $\dim \mathcal{V}^* = (\dim \mathcal{V})(\dim \mathbb{F}) = \dim \mathcal{V}$. ∎

We combine part of Theorem 71A with a converse.

•71B. PROPOSITION. *Let the linear space $\mathcal{V}$ be given. Then $\mathcal{V}^*$ is finite-dimensional if and •only if $\mathcal{V}$ is finite-dimensional.*

Proof. The "if" part follows from Theorem 71A. The "•only if" part is a special case of •Proposition 53R. For a more appealing proof of the "only if" part, we recall that there exists an injective linear mapping from $\mathcal{V}$ to $\mathcal{V}^*$ (•Corollary 61H). If $\mathcal{V}^*$ is finite-dimensional, it then follows by Proposition 53D,(a) that $\mathcal{V}$ is also finite-dimensional. ∎

From this point on we shall, in this chapter, deal almost exclusively with finite-dimensional linear spaces. A technical remark is in order here.

71C. REMARK. In the preceding sections of this chapter, some proofs depended on the •Axiom of Choice via appeals to •Theorem 15F and •Corollaries 21L, 43H, and 43M. It was noted in Remark 52H, that when restricted to finite-dimensional spaces these four results, among others, do *not* depend on the •Axiom of Choice; therefore they, together with the results in Chapter 6 whose proof relied on them, viz., •Examples 61D,(b), •Lemma 62G, •Theorem 62H, •Propositions 62O, 62S, 62U, and 64G, and • Corollaries 61H, 62I, 62J, 62M, 63E, and 63H, may, in that restricted context, be cited without the "bullet" •. ∎

For finite-dimensional spaces, many results in Chapter 6 take a simpler form.

71D. PROPOSITION. *Let the finite-dimensional linear space $\mathcal{V}$ be given.*

(a): $\mathcal{U}^{\perp}{}_{\perp} = \mathcal{U}$ *for every subspace $\mathcal{U}$ of $\mathcal{V}$ and $\mathcal{W}_{\perp}{}^{\perp} = \mathcal{W}$ for every subspace $\mathcal{W}$ of $\mathcal{V}^{*}$.*

(b): $\mathcal{A}^{\perp}{}_{\perp} = \mathrm{Lsp}\mathcal{A}$ *for every subset $\mathcal{A}$ of $\mathcal{V}$, and $\mathcal{B}_{\perp}{}^{\perp} = \mathrm{Lsp}\mathcal{B}$ for every subset $\mathcal{B}$ of $\mathcal{V}^{*}$.*

(c): *The mappings*

$$\mathcal{U} \mapsto \mathcal{U}^{\perp} \ : \ \mathrm{Subsp}(\mathcal{V}) \to \mathrm{Subsp}(\mathcal{V}^{*})$$

$$\mathcal{W} \mapsto \mathcal{W}_{\perp} \ \ \mathrm{Subsp}(\mathcal{V}^{*}) \to \mathrm{Subsp}(\mathcal{V})$$

are order-antimorphisms (with respect to inclusion), and each is the inverse of the other.

Proof. (a) and (b) follow by Theorems 62H and 62K. The mappings in (c) are antitone (Corollary 62E,(a)). By (a) each is the inverse of the other; therefore they are order-antimorphisms. ∎

71E. PROPOSITION. *Let the finite-dimensional linear space $\mathcal{V}$ be given. Then*

$$(71.1) \qquad \dim\mathcal{U} + \dim\mathcal{U}^{\perp} = \dim\mathcal{V} \quad \textit{for all subspaces } \mathcal{U} \textit{ of } \mathcal{V}$$

$$(71.2) \qquad \dim\mathcal{W} + \dim\mathcal{W}_{\perp} = \dim\mathcal{V} \quad \textit{for all subspaces } \mathcal{W} \textit{ of } \mathcal{V}^{*}$$

Proof. Let the subspace $\mathcal{U}$ of $\mathcal{V}$ be given. By Proposition 62N, $\mathcal{U}^{\perp}$ is linearly isomorphic to $(\mathcal{V}/\mathcal{U})^{*}$, and hence, by Theorem 71A and Theorem 53B, $\dim\mathcal{U}^{\perp} = \dim(\mathcal{V}/\mathcal{U})$. By Corollary 53I, $\dim\mathcal{U} + \dim(\mathcal{V}/\mathcal{U}) = \dim\mathcal{V}$, so that (71.1) holds.

Let the subspace $\mathcal{W}$ of $\mathcal{V}^{*}$ be given. We apply (71.1) to $\mathcal{U} := \mathcal{W}_{\perp}$ and find, using Proposition 71D,(a),

$$\dim\mathcal{W} + \dim\mathcal{W}_{\perp} = \dim\mathcal{W}_{\perp}{}^{\perp} + \dim\mathcal{W}_{\perp} = \dim\mathcal{V}. \ \blacksquare$$

71F. PROPOSITION. *Let the finite-dimensional linear spaces $\mathcal{V}$ and $\mathcal{W}$ be given. For every $L \in \mathrm{Lin}(\mathcal{V}, \mathcal{W})$, we have*

$$\mathrm{Null}L = (\mathrm{Rng}L^{\mathsf{T}})_{\perp} \quad \mathrm{Rng}L = (\mathrm{Null}L^{\mathsf{T}})_{\perp} \quad \mathrm{Null}L^{\mathsf{T}} = (\mathrm{Rng}L)^{\perp} \quad \mathrm{Rng}L^{\mathsf{T}} = (\mathrm{Null}L)^{\perp}.$$

Proof. The first and second inequalities follow from Corollary 62I, the third from Proposition 62F,(a), and the fourth from Corollary 62M. The latter two also follow from the former two by Proposition 71D,(a). ∎

Theorem 71A implies that every finite-dimensional linear space $\mathcal{V}$ is linearly isomorphic to its dual space, but does not provide a unique "natural" isomorphism. We shall later associate one such isomorphism with each basis of $\mathcal{V}$ (Proposition 72A,(c)).

The situation is quite different when it comes to a finite-dimensional linear space and its *second* dual space, as we now show.

71G. Theorem. *Let the finite-dimensional linear space* V *be given. Then* $\mathrm{Ev}_V \in \mathrm{Lin}(V, V^{**})$ *is invertible, i.e., a linear isomorphism.*

Proof. By Theorem 71A, V^{**} is finite-dimensional and $\dim V^{**} = \dim V^* = \dim V$. By Corollary 63E, Ev_V is injective. It follows by Corollary 53J, (i) $\Rightarrow$ (iii) that Ev_V is invertible. ∎

71H. Remarks. (a): We shall provide another type of proof of Theorem 71G later (Remark 72D).

(b): Theorem 71G encourages us to use the linear isomorphism Ev_V to identify V and V^{**}. We shall ultimately discuss this identification in Section 74, though with somewhat less than the alacrity usually displayed in this matter; we defer doing so at this point, however, since we fear the possibility of hidden notational clashes. ∎

71I. Proposition. *Let the linear spaces* V *and* W *be given, and assume that* W *is finite-dimensional. Then each of the linear mappings*

$$L \mapsto L^{\mathsf{T}} : \mathrm{Lin}(V, W) \to \mathrm{Lin}(W^*, V^*)$$

$$M \mapsto \mathrm{Ev}_W^{-1} M^{\mathsf{T}} \mathrm{Ev}_V : \mathrm{Lin}(W^*, V^*) \to \mathrm{Lin}(V, W)$$

is the inverse of the other. In particular, the linear transposition for (V, W) *is invertible.*

Proof. Using Proposition 63C, we have

$$\mathrm{Ev}_W^{-1} L^{\mathsf{TT}} \mathrm{Ev}_V = \mathrm{Ev}_W^{-1} \mathrm{Ev}_W L = L \quad \text{for all } L \in \mathrm{Lin}(V, W).$$

On the other hand, using Propositions 61C,(d) and 63D, and Corollary 63F, we have

$$
\begin{aligned}
(\mathrm{Ev}_W^{-1} M^{\mathsf{T}} \mathrm{Ev}_V)^{\mathsf{T}} &= \mathrm{Ev}_V{}^{\mathsf{T}} M^{\mathsf{TT}} (\mathrm{Ev}_W{}^{\mathsf{T}})^{-1} = \mathrm{Ev}_V{}^{\mathsf{T}} M^{\mathsf{TT}} (\mathrm{Ev}_W{}^{\mathsf{T}})^{-1} \mathrm{Ev}_W{}^{\mathsf{T}} \mathrm{Ev}_{W^*} = \\
&= \mathrm{Ev}_V{}^{\mathsf{T}} M^{\mathsf{TT}} \mathrm{Ev}_{W^*} = M \quad \text{for all } M \in \mathrm{Lin}(W^*, V^*). \quad ∎
\end{aligned}
$$

71J. Remark. Let the finite-dimensional linear spaces V and W be given. By Theorem 71A and Corollary 53O we have $\dim \mathrm{Lin}(V, W) = (\dim V)(\dim W) = (\dim W^*)(\dim V^*) = \dim \mathrm{Lin}(W^*, V^*)$. By Corollary 62J the linear mapping $(L \mapsto L^{\mathsf{T}}) : \mathrm{Lin}(V, W) \to \mathrm{Lin}(W^*, V^*)$ is injective; it follows by Corollary 53J,(i)$\Rightarrow$(iii) that it is invertible. This proof does not use Ev_V or Ev_W, but the result is both weaker than Proposition 71I (since V is also assumed to be finite-dimensional) and less explicit (since the inverse mapping is not exhibited). ∎

We may ask whether Theorem 71G and Proposition 71I are "best possible", in that the conclusion would fail to hold if the space V or the space W, respectively, were not finite-dimensional. We now show that this is indeed so.

•**71K. Proposition.** *Let the linear space* V *be given. Then* Ev_V *is invertible if and* •*only if* V *is finite- dimensional.*

Proof. The "if" part is Theorem 71G. To prove the "•only if" part, assume that $\text{Ev}_{\mathcal{V}}$ is invertible, and choose a basis $b \in \mathcal{V}^I$ of $\mathcal{V}$ (•Corollary 53M). Let $v \in \mathcal{V}$ be given. If $b_i^* v = 0$ for all $i \in I$, it follows by Corollary 64C that $v = 0$. Thus $(\text{Rng}b^*)_\perp = \{0\}$.

Since $\text{Ev}_{\mathcal{V}}$ is surjective, use of Proposition 63G yields

$$(\text{Rng}b^*)^\perp = \text{Ev}_{\mathcal{V}>}(\text{Ev}_{\mathcal{V}}{}^<((\text{Rng}b^*)^\perp)) = \text{Ev}_{\mathcal{V}>}((\text{Rng}b^*)_\perp) = \text{Ev}_{\mathcal{V}>}(\{0\}) = \{0\},$$

and therefore, by •Theorem 62H,(b),

$$\text{Lsp Rng}b^* = (\text{Rng}b^*)^\perp{}_\perp = \{0\}_\perp = \mathcal{V}^*.$$

The linearly independent family b^* (Proposition 64A) thus spans $\mathcal{V}^*$, and is therefore a basis of $\mathcal{V}^*$. By Proposition 64F, $\mathcal{V}$ is finite- dimensional. ∎

•**71L. PROPOSITION.** *Let the linear spaces $\mathcal{V}$ and $\mathcal{W}$ be given. The linear transposition for $(\mathcal{V}, \mathcal{W})$ is invertible if and •only if $\mathcal{W}$ is finite-dimensional or $\mathcal{V}$ is a zero-space.*

Proof. If $\mathcal{V}$ is a zero-space, then $\mathcal{V}^*$, $\text{Lin}(\mathcal{V}, \mathcal{W})$, $\text{Lin}(\mathcal{W}^*, \mathcal{V}^*)$ are also zero-spaces and the linear transposition is trivially invertible. If $\mathcal{W}$ is finite-dimensional, the linear transposition is invertible by Proposition 71I. This completes the proof of the "if" part.

To prove the "•only if" part, assume that $\mathcal{V} \neq \{0\}$ and that the linear transposition is invertible. Choose $u \in \mathcal{V}^\times$. By •Lemma 62G applied to $\mathcal{U} := \{0\}$ we may choose $\lambda \in \mathcal{V}^*$ such that $\lambda u = 1$.

Now let $\Pi \in \mathcal{W}^{**}$ be given. Then $\lambda \otimes \Pi \in \text{Lin}(\mathcal{W}^*, \mathcal{V}^*)$, and therefore $\lambda \otimes \Pi = L^\mathsf{T}$ for a suitable $L \in \text{Lin}(\mathcal{V}, \mathcal{W})$. We then have

$$(\text{Ev}_{\mathcal{W}}(Lu))\mu = \mu Lu = (L^\mathsf{T}\mu)u = ((\lambda \otimes \Pi)\mu)u = (\Pi\mu)(\lambda u) = \Pi\mu \quad \text{for all } \mu \in \mathcal{W}^*.$$

Therefore $\Pi = \text{Ev}_{\mathcal{W}}(Lu)$. Since $\Pi \in \mathcal{W}^{**}$ was arbitrary, it follows that $\text{Ev}_{\mathcal{W}}$ is surjective, hence invertible (•Corollary 63E). By •Proposition 71K, $\mathcal{W}$ is finite-dimensional.

▲   ∎

72. Dual bases and matrices

Let the finite-dimensional linear space $\mathcal{V}$ and the basis $b \in \mathcal{V}^I$ of $\mathcal{V}$ be given. Then b^* is a basis of $\mathcal{V}^*$ (Proposition 64F), the *basis (of $\mathcal{V}^*$) biorthogonal* to b; when b is understood, b^* is called the **dual basis** of $\mathcal{V}^*$ for short.

72A. PROPOSITION. *Let the finite-dimensional linear space $\mathcal{V}$ and the basis $b \in \mathcal{V}^I$ of $\mathcal{V}$ be given.*

(a): *For every linear space $\mathcal{W}$, family $y \in \mathcal{W}^I$, and linear mapping $L \in \mathrm{Lin}(\mathcal{V}, \mathcal{W})$ we have $Lb = y$ if and only if $L = \sum_{i \in I} y_i \otimes b_i^*$.*

(b): $\displaystyle\sum_{i \in I} b_i \otimes b_i^* = 1_{\mathcal{V}}$.

(c): $L := \displaystyle\sum_{i \in I} b_i^* \otimes b_i^*$ *is the only solution to the problem*

$$? L \in \mathrm{Lin}(\mathcal{V}, \mathcal{V}^*), \quad Lb = b^*,$$

and it is a linear isomorphism.

Proof. (a) follows from Corollary 64E, using the fact that I is finite. (b) and (c) are special cases of (a); in (c), L is invertible because b^* is a basis of $\mathcal{V}^*$ (Proposition 43D). ∎

72B. COROLLARY. *Let the linear spaces $\mathcal{V}$ and $\mathcal{W}$ be given. If either $\mathcal{V}$ or $\mathcal{W}$ is finite-dimensional, then the set $\{w \otimes \lambda \mid w \in \mathcal{W},\ \lambda \in \mathcal{V}^*\}$ spans $\mathrm{Lin}(\mathcal{V}, \mathcal{W})$.*

Proof. If $\mathcal{V}$ is finite-dimensional, choose a basis $b \in \mathcal{V}^I$ of $\mathcal{V}$ (Corollary 53C). By Proposition 72A,(a) we have $L = \displaystyle\sum_{i \in I} (Lb_i) \otimes b_i^*$ for all $L \in \mathrm{Lin}(\mathcal{V}, \mathcal{W})$. If $\mathcal{W}$ is finite-dimensional, choose a basis $c \in \mathcal{W}^J$ of $\mathcal{W}$. By Proposition 72A,(b), $L = \displaystyle\sum_{j \in J} c_j \otimes c_j^*)L = (\sum_{j \in J} c_j \otimes (c_j^* L)$ for all $L \in \mathrm{Lin}(\mathcal{V}, \mathcal{W})$. In either case, $L \in \mathrm{Lsp}\{w \otimes \lambda \mid w \in \mathcal{W},\ \lambda \in \mathcal{V}^*\}$ for all $L \in \mathrm{Lin}(\mathcal{V}, \mathcal{W})$. ∎

▼ Is Corollary 72B the best possible result? The following proposition shows that it is.

72C. PROPOSITION. *Let the linear spaces $\mathcal{V}$ and $\mathcal{W}$ be given. Then* $\mathrm{Lsp}\{w \otimes \lambda \mid w \in \mathcal{W},\ \lambda \in \mathcal{V}^*\} = \{L \in \mathrm{Lin}(\mathcal{V}, \mathcal{W}) \mid \mathrm{Rng}L$ *is finite-dimensional*$\}$.

Proof. 1. Let $L \in \mathrm{Lsp}\{w \otimes \lambda \mid w \in \mathcal{W},\ \lambda \in \mathcal{V}^*\}$ be given. By Corollary 13O, we may choose a finite subset $\mathcal{F}$ of $\mathcal{W} \times \mathcal{V}^*$ such that $L \in \mathrm{Lsp}\{w \otimes \lambda \mid (w, \lambda) \in \mathcal{F}\}$. By Theorem 41C we may then choose $a \in \mathbb{F}^{\mathcal{F}}$ such that $L = \displaystyle\sum_{(w,\lambda) \in \mathcal{F}} a_{(w,\lambda)} w \otimes \lambda$. But then

$$Lv = \sum_{(w,\lambda) \in \mathcal{F}} a_{(w,\lambda)} w \otimes \lambda v = \sum_{(w,\lambda) \in \mathcal{F}} (a_{(w,\lambda)} \lambda v) w \quad \text{for all } v \in \mathcal{V},$$

and therefore $\mathrm{Rng}L$ is a subspace of the finite-dimensional subspace

$\mathrm{Lsp}\{w \in \mathcal{W} \mid (w, \lambda) \in \mathcal{F}\}$ of $\mathcal{W}$, so that $\mathrm{Rng}L$ is itself finite-dimensional.

2. Let $L \in \mathrm{Lin}(\mathcal{V}, \mathcal{W})$ be given, and assume that $\mathcal{X} := \mathrm{Rng}L$ is finite-dimensional. By Corollary 72B, $L' := L|^{\mathcal{X}} \in \mathrm{Lsp}\{w \otimes^{\mathcal{X}} \lambda \mid w \in \mathcal{X},\ \lambda \in \mathcal{V}^*\}$. Then

$$L = 1_{\mathcal{X} \subset \mathcal{W}} L' \in \mathrm{Lsp}\{1_{\mathcal{X} \subset \mathcal{W}}(w \otimes^{\mathcal{X}})\lambda \mid w \in \mathcal{X},\ \lambda \in \mathcal{V}^*\} =$$
$$= \mathrm{Lsp}\{w \otimes^{\mathcal{W}} \lambda \mid w \in \mathcal{X},\ \lambda \in \mathcal{V}^*\} \subset \mathrm{Lsp}\{w \otimes \lambda \mid w \in \mathcal{W},\ \lambda \in \mathcal{V}^*\}. \quad \blacksquare$$

72D. COROLLARY. *Let the linear space $\mathcal{V}$ be given. The set $\{v \otimes \lambda \mid v \in \mathcal{V},\ \lambda \in \mathcal{V}^*\}$ spans $\mathrm{Lin}\mathcal{V}$ if and only if $\mathcal{V}$ is finite-dimensional.*

Proof. Apply Proposition 72C to $L := 1_{\mathcal{V}}$. $\blacksquare$

•**72E. COROLLARY.** *Let the linear spaces $\mathcal{V}$ and $\mathcal{W}$ be given. Then the set $\{w \otimes \lambda \mid w \in \mathcal{W},\ \lambda \in \mathcal{V}^*\}$ spans $\mathrm{Lin}(\mathcal{V}, \mathcal{W})$ if and* •*only if either $\mathcal{V}$ or $\mathcal{W}$ is finite-dimensional.*

Proof. The "if" part is Corollary 72B. We prove the "•only if" part by contraposition. We assume that $\mathcal{V}$ and $\mathcal{W}$ are infinite-dimensional, and choose basis-sets $\mathcal{B}$ of $\mathcal{V}$ and $\mathcal{C}$ of $\mathcal{W}$, respectively (•Corollary 53O). Since $\mathcal{B}$ and $\mathcal{C}$ are both infinite, we may choose injective sequences $b \in \mathcal{B}^{\mathbb{N}}$ and $c \in \mathcal{C}^{\mathbb{N}}$ (*Basic Language,* •Theorem 121V). There exists exactly one $L \in \mathrm{Lin}(\mathcal{V}, \mathcal{W})$ such that $Lb = c$ and $Lv = 0$ for all $v \in \mathcal{B}\backslash\mathrm{Rng}b$ (Theorem 43F,(c)). But then $\mathrm{Rng}c$ is infinite and linearly independent, and $\mathrm{Rng}c \subset \mathrm{Rng}L$; therefore $\mathrm{Rng}L$ is not finite-dimensional (Theorem 52A,(a)). By Proposition 72C we have $L \notin \mathrm{Lsp}\{w \otimes \lambda \mid w \in \mathcal{W},\ \lambda \in \mathcal{V}^*\}$. $\blacksquare$

We next consider dual bases of dual bases.

72F. PROPOSITION. *Let the finite-dimensional linear space $\mathcal{V}$ and the basis $b \in \mathcal{V}^I$ be given. Then*

$$(72.1) \qquad\qquad\qquad \mathrm{Ev}_{\mathcal{V}}b = b^{**}$$

$$(72.2) \qquad\qquad \mathrm{Ev}_{\mathcal{V}} = \Big(\sum_{i \in I} b_i^{**} \otimes b_i^{**}\Big)\Big(\sum_{i \in I} b_i^* \otimes b_i^*\Big).$$

Proof. We have

$$(\mathrm{Ev}_{\mathcal{V}}b_i)b_j^* = b_j^* b_i = \delta_{j,i}^I \quad \text{for all } i, j \in I.$$

Thus $\mathrm{Ev}_{\mathcal{V}}b$ is biorthogonal to the basis b^* of $\mathcal{V}^*$. By Theorem 64B, $\mathrm{Ev}_{\mathcal{V}}b = b^{**}$.

By Proposition 72A,(c) we have, using (72.1),

$$\Big(\sum_{i \in I} b_i^{**} \otimes b_i^{**}\Big)\Big(\sum_{i \in I} b_i^* \otimes b_i^*\Big)b_j = \Big(\sum_{i \in I} b_i^{**} \otimes b_i^{**}\Big)b_j^* = b_j^{**} = \mathrm{Ev}_{\mathcal{V}}b_j \quad \text{for all } j \in I.$$

Since b is a basis of $\mathcal{V}$, (72.2) follows by Theorem 43F,(c). A direct computation, with application of the distributive law to the right-hand side of (72.2), can also be carried out to prove the equality. $\blacksquare$

72G. REMARK. Given a finite-dimensional linear space V, we may choose a basis $b \in V^I$ of V. Formula (72.2) then exhibits Ev_V as the composite of two linear isomorphisms, and hence Ev_V is itself a linear isomorphism. Alternatively, (72.1) shows that Ev_V is invertible, since b^{**} is a basis of V^{**}. We thus have two other proofs of Theorem 71G. ∎

We next examine the relationship between bases and their dual bases on the one hand, and the matrices of linear mappings and of their transposes on the other. We recall from *Basic Language*, Section 42 that the **transpose of** a $J \times I$-matrix M is the $I \times J$-matrix $M^\top$ that satisfies $M^\top_{i,j} := M_{j,i}$ for all $(i,j) \in I \times J$.

72H. PROPOSITION. *Let the finite-dimensional linear spaces V and W and the bases $b \in V^I$ of V and $c \in W^J$ of W be given. For every $L \in \mathrm{Lin}(V, W)$ we have*

$$(72.3) \qquad [L]^{c,b}_{j,i} = c^*_j L b_i \quad \text{for all } (j,i) \in J \times I$$

$$(72.4) \qquad L = \sum_{(j,i) \in J \times I} [L]^{c,b}_{j,i} c_j \otimes b^*_i$$

$$(72.5) \qquad [L^\top]^{b^*,c^*} = ([L])^{c,b})^\top.$$

Proof. From (53.4) we obtain

$$c^*_j L b_i = \sum_{k \in J} [L]^{c,b}_{k,i} c^*_j c_k = \sum_{k \in J} [L]^{c,b}_{k,i} \delta^J_{j,k} = [L]^{c,b}_{j,i} \quad \text{for all } (j,i) \in J \times I;$$

combining the same formula with Proposition 72A,(a) we obtain

$$L = \sum_{i \in I} L b_i \otimes b^*_i = \sum_{i \in I} \sum_{j \in J} [L]^{c,b}_{j,i} c_j \otimes b^*_i = \sum_{(j,i) \in J \times I} [L]^{c,b}_{j,i} c_j \otimes b^*_i.$$

Thus (72.3) and (72.4) hold. Finally, we apply (72.3) twice, once in the given form and once with L, b, c replaced by $L^\top$, c^*, b^*, respectively, and find using (72.1),

$$[L^\top]^{b^*,c^*}_{i,j} = b^{**}_i L^\top c^*_j = (\mathrm{Ev}_V b_i) L^\top c^*_j = (L^\top c^*_j) b_i = c^*_j L b_i = [L]^{c,b}_{j,i}$$
$$\text{for all } (i,j) \in I \times J.$$

Therefore (72.5) also holds. ∎

The validity of (72.5) first suggested the term "transpose of L" and the notation $L^\top$ for the object defined in Section 61 for a linear mapping L, even when the domain or the codomain of L is not finite-dimensional.

72I. COROLLARY. *Let the finite-dimensional linear spaces V and W and the bases $b \in V^I$ of V and $c \in W^J$ of W be given. Then $(c_j \otimes b^*_i \mid (j,i) \in J \times I)$ is a basis of $\mathrm{Lin}(V, W)$.*

Proof. By Proposition 72H (formula (72.4)) we have, for every pair $(j, i) \in J \times I$,

$$\forall L \in \mathrm{Lin}(\mathcal{V}, \mathcal{W}), \quad [L]^{c,b} = \delta_{(j,i)}^{J \times I} \quad \Leftrightarrow \quad L = c_j \otimes b_i^*.$$

It follows that $(c_j \otimes b_i^* \mid (j, i) \in J \times I)$ is the family of images of the terms of the basis $(\delta_{(j,i)}^{J \times I} \mid (j, i) \in J \times I)$ of $\mathbb{F}^{J \times I}$ under the inverse of the linear isomorphism $(L \mapsto [L]^{c,b}) : \mathrm{Lin}(\mathcal{V}, \mathcal{W}) \to \mathbb{F}^{J \times I}$ (Proposition 53N). Therefore $(c_j \otimes b_i^* \mid (j, i) \in J \times I)$ is a basis of $\mathrm{Lin}(\mathcal{V}, \mathcal{W})$ (Proposition 43D). ∎

72J. Remark. It is not hard to show, with the help of Corollary 72B, that the conclusion of Corollary 72I remains valid if it is merely assumed that $\mathcal{V}$ or $\mathcal{W}$ is finite-dimensional. ∎

73. The trace form

In this section we intend to examine the structure of the dual space of $\mathrm{Lin}(\mathcal{V}, \mathcal{W})$ for given finite-dimensional linear spaces $\mathcal{V}$ and $\mathcal{W}$. In particular, we shall introduce, for every given finite-dimensional linear space $\mathcal{V}$, a distinguished linear form on $\mathrm{Lin}\mathcal{V}$, called the *trace form of $\mathcal{V}$*.

We record a recurring proof pattern in the form of a lemma.

73A. Lemma. *Let the linear spaces $\mathcal{V}$, $\mathcal{W}$, $\mathcal{X}$ be given, and assume that either $\mathcal{V}$ or $\mathcal{W}$ is finite-dimensional. Let $L, M \in \mathrm{Lin}(\mathrm{Lin}(\mathcal{V}, \mathcal{W}), \mathcal{X})$ be given, and assume that*

$$L(w \otimes \lambda) = M(w \otimes \lambda) \quad for\ all\ w \in \mathcal{W}\ and\ \lambda \in \mathcal{V}^*.$$

Then $L = M$.

Proof. By Corollary 72B, the set $\{w \otimes \lambda \mid w \in \mathcal{W},\ \lambda \in \mathcal{V}^*\}$ spans $\mathrm{Lin}(\mathcal{V}, \mathcal{W})$. The conclusion then follows by Theorem 43F,(b). ∎

73B. Proposition. *Let the linear spaces $\mathcal{V}$ and $\mathcal{W}$ be given. There exists exactly one mapping $\gamma : (\mathrm{Lin}(\mathcal{V}, \mathcal{W}))^* \to \mathrm{Lin}(\mathcal{V}^*, \mathcal{W}^*)$ such that*

$$(73.1)\quad ((\gamma(\Lambda))\lambda)w = \Lambda(w \otimes \lambda) \quad for\ all\ w \in \mathcal{W}\ and\ \lambda \in \mathcal{V}^*\ and\ \Lambda \in (\mathrm{Lin}(\mathcal{V}, \mathcal{W}))^*.$$

This mapping is linear. If $\mathcal{V}$ or $\mathcal{W}$ is finite-dimensional, γ is injective; if both $\mathcal{V}$ and $\mathcal{W}$ are finite-dimensional, γ is invertible.

Proof. For every $\Lambda \in (\mathrm{Lin}(\mathcal{V}, \mathcal{W}))^*$ and every $\lambda \in \mathcal{V}^*$ the composite

$$w \mapsto \Lambda(w \otimes \lambda) : \mathcal{W} \to \mathbb{F}$$

of the linear mappings $(w \mapsto w \otimes \lambda) : \mathcal{W} \to \mathrm{Lin}(\mathcal{V}, \mathcal{W})$ and Λ is linear, hence a member of $\mathcal{W}^*$. This defines a mapping

$$\lambda \mapsto (w \mapsto \Lambda(w \otimes \lambda)) : \mathcal{V}^* \to \mathcal{W}^*$$

for each $\Lambda \in (\mathrm{Lin}(\mathcal{V}, \mathcal{W}))^*$; this mapping is itself obviously linear, hence a member of $\mathrm{Lin}(\mathcal{V}^*, \mathcal{W}^*)$. This establishes the existence and uniqueness of the mapping $\gamma : (\mathrm{Lin}(\mathcal{V}, \mathcal{W}))^* \to \mathrm{Lin}(\mathcal{V}^*, \mathcal{W}^*)$ satisfying (73.1). The linearity of γ is obvious.

2. Assume that $\mathcal{V}$ or $\mathcal{W}$ is finite-dimensional. Let $\Lambda \in \mathrm{Null}\gamma$ be given. From (73.1) we have $\Lambda(w \otimes \lambda) = 0$ for all $w \in \mathcal{W}$ and $\lambda \in \mathcal{V}^*$. By Lemma 73A, $\Lambda = 0$. Thus $\mathrm{Null}\gamma = \{0\}$, and γ is injective.

3. Assume that both $\mathcal{V}$ and $\mathcal{W}$ are finite-dimensional. Then $(\mathrm{Lin}(\mathcal{V}, \mathcal{W}))^*$ and $\mathrm{Lin}(\mathcal{V}^*, \mathcal{W}^*)$ are also finite-dimensional, and, by Theorem 71A and Corollary 53O,

$$\dim(\mathrm{Lin}(\mathcal{V}, \mathcal{W}))^* = \dim\ \mathrm{Lin}(\mathcal{V}, \mathcal{W}) = (\dim\mathcal{V})(\dim\mathcal{W}) = (\dim\mathcal{V}^*)(\dim(\mathcal{W}^*) =$$
$$= \dim\ \mathrm{Lin}(\mathcal{V}^*, \mathcal{W}^*).$$

Since γ is injective, it follows by Corollary 53J that γ is invertible. ∎

We shall use the temporary notation $\Gamma_{V,W}$ for the unique mapping γ satisfying (73.1). Thus $\Gamma_{V,W} \in \mathrm{Lin}((\mathrm{Lin}(V,W))^*, \mathrm{Lin}(V^*,W^*))$, and

$$(73.2) \quad ((\Gamma_{V,W}\Lambda)\lambda)w = \Lambda(w \otimes \lambda) \quad \text{for all } w \in W \text{ and } \lambda \in V^* \text{ and } \Lambda \in (\mathrm{Lin}(V,W))^*.$$

Let the finite-dimensional linear space V be given. Then $\Gamma_{V,V}$ is invertible, and we define

$$(73.3) \qquad\qquad \mathrm{tr}_V := \Gamma_{V,V}{}^{-1}1_{V^*} \in (\mathrm{Lin}V)^*;$$

this linear form on $\mathrm{Lin}V$ is called the **trace form of** V. Its significance stems from the following characterization.

73C. Theorem. *Let the finite-dimensional linear space V be given. Then the problem*

$$(73.4) \qquad ?T \in (\mathrm{Lin}V)^*, \quad (\forall v \in V, \ \forall \lambda \in V^*, \quad T(v \otimes \lambda) = \lambda v)$$

has exactly one solution, namely tr_V.

Proof. Using (73.2) and (73.3), we find

$$\mathrm{tr}_V(v \otimes \lambda) = ((\Gamma_{V,V}\mathrm{tr}_V)\lambda)v = (1_{V^*}\lambda)v = \lambda v \quad \text{for all } v \in V \text{ and } \lambda \in V^*.$$

This shows that $\mathrm{tr}V$ is a solution of (73.4); Lemma 73A shows that it is the only solution. ∎

We record the formula, valid for every finite-dimensional space V,

$$(73.5) \qquad\qquad \mathrm{tr}_V(v \otimes \lambda) = \lambda v \quad \text{for all } v \in V \text{ and } \lambda \in V^*.$$

For every $L \in \mathrm{Lin}V$, we may write $\mathrm{tr}L$ for $\mathrm{tr}_V L$ without danger of confusion; $\mathrm{tr}L$ is called the **trace of** L.

We examine other properties of the trace forms.

73D. Proposition. *Let the finite-dimensional linear spaces V and W be given. Then*

$$(73.6) \qquad \mathrm{tr}_V(ML) = \mathrm{tr}_W(LM) \quad \text{for all } L \in \mathrm{Lin}(V,W) \text{ and } M \in \mathrm{Lin}(W,V).$$

Proof. Let $M \in \mathrm{Lin}(W,V)$ be given. Then

$$\mathrm{tr}_V(M(w \otimes \lambda)) = \mathrm{tr}_V((Mw) \otimes \lambda) = \lambda Mw = \mathrm{tr}_W(w \otimes (\lambda M)) = \mathrm{tr}_W((w \otimes \lambda)M)$$
$$\text{for all } w \in W \text{ and } \lambda \in V^*.$$

We consider the linear forms $(L \mapsto \mathrm{tr}_\mathcal{V}(ML)), (L \mapsto \mathrm{tr}_\mathcal{W}(LM)) \in (\mathrm{Lin}(\mathcal{V}, \mathcal{W}))^*$; the preceeding computation and Lemma 73A show that these linear forms are equal. ∎

73E. Corollary. *Let the finite-dimensional linear spaces $\mathcal{V}$ and $\mathcal{W}$ and the linear isomorphism $A \in \mathrm{Lis}(\mathcal{V}, \mathcal{W})$ be given. Then*

$$\mathrm{tr}_\mathcal{W}(ALA^{-1}) = \mathrm{tr}_\mathcal{V} L \quad \textit{for all } L \in \mathrm{Lin}\mathcal{V}.$$

The following result shows that (73.6) is a sort of characterization (up to "scalar multiples") of the trace forms.

73F. Proposition. *Let the finite-dimensional linear spaces $\mathcal{V}$ and $\mathcal{W}$ be given, and assume that neither is a zero-space. Let $\Sigma \in (\mathrm{Lin}\mathcal{V})^*$ and $\Sigma' \in (\mathrm{Lin}\mathcal{W})^*$ be given. Then*

$$(73.7) \qquad \Sigma(ML) = \Sigma'(LM) \quad \textit{for all } L \in \mathrm{Lin}(\mathcal{V}, \mathcal{W}) \textit{ and } M \in \mathrm{Lin}(\mathcal{W}, \mathcal{V})$$

if and only if $\Sigma = s\,\mathrm{tr}_\mathcal{V}$ and $\Sigma' = s\,\mathrm{tr}_\mathcal{W}$ for some $s \in \mathbb{F}$.

Proof. The "if" part follows by Proposition 73D. To prove the "only if" part, we assume that (73.7) holds. We choose $u \in \mathcal{V}^\times$ and $x \in \mathcal{W}^\times$. We may then choose $\beta \in \mathcal{V}^*$ and $\gamma \in \mathcal{W}^*$ such that $\beta u = 1 = \gamma x$, by Lemma 62G (cf. Remark 71C).

Using (73.7) and (73.5), we have

$$\Sigma(v \otimes \lambda) = \Sigma((\gamma x)(v \otimes \lambda)) = \Sigma((v \otimes \gamma)(x \otimes \lambda)) = \Sigma'((x \otimes \lambda)(v \otimes \gamma)) =$$
$$= \Sigma'((\lambda v)(x \otimes \gamma)) = (\lambda v)\Sigma'(x \otimes \gamma) = (\Sigma'(x \otimes \gamma))\mathrm{tr}_\mathcal{V}(v \otimes \lambda) \quad \textit{for all } v \in \mathcal{V} \textit{ and } \lambda \in \mathcal{V}^*.$$

By Lemma 73A we then conclude that

$$(73.8) \qquad\qquad\qquad \Sigma = (\Sigma'(x \otimes \gamma))\mathrm{tr}_\mathcal{V}.$$

Similarly, we find that

$$\Sigma'(w \otimes \mu) = (\Sigma(u \otimes \beta))\mathrm{tr}_\mathcal{W}(w \otimes \mu) \quad \textit{for all } w \in \mathcal{W} \textit{ and } \mu \in \mathcal{W}^*,$$

so that

$$(73.9) \qquad\qquad\qquad \Sigma' = (\Sigma(u \otimes \beta))\mathrm{tr}_\mathcal{W}.$$

Finally, (73.8) yields

$$\Sigma(u \otimes \beta) = (\Sigma'(x \otimes \gamma))\mathrm{tr}_\mathcal{V}(u \otimes \beta) = (\beta u)\Sigma'(x \otimes \gamma) = \Sigma'(x \otimes \gamma).$$

Combination of this with (73.8) and (73.9) shows that $\Sigma = s\,\mathrm{tr}_\mathcal{V}$ and $\Sigma' = s\,\mathrm{tr}_\mathcal{W}$, with $s := \Sigma(u \otimes \beta)$. ∎

73G. Corollary. *Let the finite-dimensional linear space* $\mathcal{V}$ *and the linear form* $T \in (\text{Lin}\mathcal{V})^*$ *be given. Then*

(73.10) $T(ML) = T(LM)$ *for all* $L, M \in \text{Lin}\mathcal{V}$

if and only if $T = s\,\text{tr}_\mathcal{V}$ *for some* $s \in \mathbb{F}$.

73H. Proposition. *Let the finite-dimensional linear space* $\mathcal{V}$ *be given. Then*

$$\text{tr}_{\mathcal{V}^*}(L^\top) = \text{tr}_\mathcal{V} L   \text{ for all } L \in \text{Lin}\mathcal{V}.$$

Proof. By (61.2) and (63.2) together with (73.5) we have

$$\text{tr}_{\mathcal{V}^*}((v \otimes \lambda)^\top) = \text{tr}_{\mathcal{V}^*}(\lambda^\top(v\otimes)^\top) = \text{tr}_{\mathcal{V}^*}(\lambda \otimes (\text{Ev}_\mathcal{V} v)) = (\text{Ev}_\mathcal{V} v)\lambda = \lambda v =$$
$$= \text{tr}_\mathcal{V}(v \otimes \lambda) \qquad\qquad \text{for all } v \in \mathcal{V} \text{ and all } \lambda \in \mathcal{V}^*$$

We may then apply Lemma 73A to the linear forms $\text{tr}_\mathcal{V}$ and $(L \mapsto \text{tr}_{\mathcal{V}^*}(L^\top)) \in (\text{Lin}\mathcal{V})^*$, and conclude that they are equal. ∎

Proposition. *Let the finite-dimensional linear space* $\mathcal{V}$ *be given. Then* $\text{tr}\, 1_\mathcal{V} = (\dim\mathcal{V})1$ *(the* $(\dim\mathcal{V})$*th natural multiple of the unity* 1 *of* $\mathbb{F}$*)*.

Proof. Choose a basis $b \in \mathcal{V}^I$ of $\mathcal{V}$. Then $\#I = \dim\mathcal{V}$ (Corollary 53C). By Proposition 72A,(b) and (73.5) we have

$$\text{tr}\, 1_\mathcal{V} = \text{tr}\sum_{i \in I} b_i \otimes b_i^* = \sum_{i \in I}\text{tr}(b_i \otimes b_i^*) = \sum_{i \in i} b_i^* b_i = \sum_{i \in I} 1 = (\#I)1 = (\dim\mathcal{V})1. ∎$$

73J. Corollary. *Let the finite-dimensional linear space* $\mathcal{V}$ *be given, and assume that either* $\dim\mathcal{V} = 0$ *or* $(\dim\mathcal{V})1 \neq 0$. *Then a linear form* $T \in (\text{Lin}\mathcal{V})^*$ *is the trace form* $\text{tr}_\mathcal{V}$ *if and only if* (73.10) *holds and* $T1_\mathcal{V} = (\dim\mathcal{V})1$.

Proof. Corollary 73G and Proposition 73I. ∎

73K. Remark. The assumption that either $\dim\mathcal{V} = 0$ or $(\dim\mathcal{V})1 \neq 0$ is certainly satisfied if the field $\mathbb{F}$ is $\mathbb{R}$ or $\mathbb{Q}$, regardless of what the dimension of $\mathcal{V}$ is. ∎

73L. Corollary. *Let the finite-dimensional linear space* $\mathcal{V}$ *and the idempotent linear mapping* $E \in \text{Lin}\mathcal{V}$ *be given. Then* $\text{tr}E = (\dim \text{Rng}E)1$.

Proof. Set $\mathcal{U} := \text{Rng}E$. Since E is idempotent, $\mathcal{U}$ is the set of fixed points of E. Thus $E|_\mathcal{U}^\mathcal{U} = 1_\mathcal{U}$. By Propositions 73D and 73I, $\text{tr}_\mathcal{V} E = \text{tr}_\mathcal{V}(1_{\mathcal{U}\subset\mathcal{V}}(E|^\mathcal{U})) = \text{tr}_\mathcal{U}((E|^\mathcal{U})1_{\mathcal{U}\subset\mathcal{V}}) = \text{tr}_\mathcal{U}(E|_\mathcal{U}^\mathcal{U}) = \text{tr}_\mathcal{U} 1_\mathcal{U} = (\dim\mathcal{U})1.$ ∎

We now show how to compute the trace of a member of $\text{Lin}\mathcal{V}$ when its (square) matrix with respect to a basis of $\mathcal{V}$ is known.

73M. Proposition. *Let the finite-dimensional linear space* $\mathcal{V}$ *and the basis* $b \in \mathcal{V}^I$ *of* $\mathcal{V}$ *be given. Then*

(73.11) $\text{tr}L = \sum_{i \in I}[L]_{i,i}^b$ *for all* $L \in \text{Lin}\mathcal{V}$.

Proof. By Proposition 72H (formula (72.4)) and (73.5) we have, for every $L \in \mathrm{Lin}\mathcal{V}$,

$$\mathrm{tr}L = \mathrm{tr}\sum_{(j,i)\in I\times I} [L]^b_{j,i} b_j \otimes b^*_i = \sum_{(j,i)\in I\times I} [L]^b_{j,i}\mathrm{tr}(b_j \otimes b^*_j) = \sum_{(j,i)\in I\times I} [L]^b_{j,i} b^*_i b_j =$$

$$= \sum_{i\in I}\sum_{j\in J} [L]^b_{j,i}\delta^I_{i,j} = \sum_{i\in I} [L]^b_{i,i}. \quad \blacksquare$$

73N. REMARK. Let the finite set I and the $I \times I$-matrix $M \in \mathbb{F}^{I\times I}$ be given. Under the identification described in Section 51, M is also a member of $\mathrm{Lin}\mathbb{F}^I$. In this capacity, its matrix with respect to the Kronecker basis is precisely the *matrix* M. Thus (73.11) yields

(73.12) $$\mathrm{tr}M = \sum_{i\in I} M_{i,i}.$$

It is customary to use (73.12) as a definition of the **trace of** the *matrix* M, and then the identification just referred to produces no notational clash. With this definition, formula (73.11) for every given finite-dimensional linear space $\mathcal{V}$ and given basis $b \in \mathcal{V}^I$ of $\mathcal{V}$ may be restated as follows:

(73.13) $$\mathrm{tr}L = \mathrm{tr}[L]^b \quad \text{for all } L \in \mathrm{Lin}\mathcal{V}. \quad \blacksquare$$

We finally return to the matter of the structure of $(\mathrm{Lin}(\mathcal{V},\mathcal{W}))^*$ for given finite-dimensional linear spaces $\mathcal{V}$ and $\mathcal{W}$. We observe that for every $M \in \mathrm{Lin}(\mathcal{W},\mathcal{V})$ the mapping

$$L \mapsto \mathrm{tr}_{\mathcal{V}}(ML) : \mathrm{Lin}(\mathcal{V},\mathcal{W}) \to \mathbb{F}$$

is linear, hence a member of $(\mathrm{Lin}(\mathcal{V},\mathcal{W}))^*$. We shall show that, conversely, *every* linear form on $\mathrm{Lin}(\mathcal{V},\mathcal{W})$ is obtained in this way from precisely one $M \in \mathrm{Lin}(\mathcal{W},\mathcal{V})$. We shall make the invertible mappings involved as explicit as possible. The preceding observation establishes the existence of precisely one (obviously linear) mapping $\Phi_{\mathcal{V},\mathcal{W}} : \mathrm{Lin}(\mathcal{W},\mathcal{V}) \to (\mathrm{Lin}(\mathcal{V},\mathcal{W}))^*$ such that

(73.14) $\quad (\Phi_{\mathcal{V},\mathcal{W}}M)L = \mathrm{tr}_{\mathcal{V}}(ML) \quad$ for all $L \in \mathrm{Lin}(\mathcal{V},\mathcal{W})$ and $M \in \mathrm{Lin}(\mathcal{W},\mathcal{V})$.

73O. THEOREM. *Let the finite-dimensional linear spaces $\mathcal{V}$ and $\mathcal{W}$ be given. Then for every $\Lambda \in (\mathrm{Lin}(\mathcal{V},\mathcal{W}))^*$ there exists exactly one $M \in \mathrm{Lin}(\mathcal{W},\mathcal{V})$ such that $\Lambda L = \mathrm{tr}_{\mathcal{V}}(ML)$ for all $L \in \mathrm{Lin}(\mathcal{V},\mathcal{W})$. More precisely, the linear mapping $\Phi_{\mathcal{V},\mathcal{W}}$ is invertible. Even more precisely, the composite $\Gamma_{\mathcal{V},\mathcal{W}}\Phi_{\mathcal{V},\mathcal{W}} : \mathrm{Lin}(\mathcal{W},\mathcal{V}) \to \mathrm{Lin}(\mathcal{V}^*,\mathcal{W}^*)$ is the linear transposition for $(\mathcal{W},\mathcal{V})$.*

Proof. Let $M \in \mathrm{Lin}(\mathcal{W}, \mathcal{V})$ be given. By (73.2), (73.14), and (73.5) we have

$$((\Gamma_{\mathcal{V},\mathcal{W}} \Phi_{\mathcal{V},\mathcal{W}} M)\lambda)w = (\Phi_{\mathcal{V},\mathcal{W}} M)(w \otimes \lambda) = \mathrm{tr}_{\mathcal{V}}(Mw \otimes \lambda) = \lambda Mw$$
$$\text{for all } w \in \mathcal{W} \text{ and } \lambda \in \mathcal{V}^*;$$

hence $(\Gamma_{\mathcal{V},\mathcal{W}} \Phi_{\mathcal{V},\mathcal{W}} M)\lambda = \lambda M = M^{\top}\lambda$ for all $\lambda \in \mathcal{V}^*$, i.e., $\Gamma_{\mathcal{V},\mathcal{W}} \Phi_{\mathcal{V},\mathcal{W}} M = M^{\top}$, as claimed.

Since both $\Gamma_{\mathcal{V},\mathcal{W}}$ and the linear transposition are invertible (Propositions 73B and 71I), it follows that $\Phi_{\mathcal{V},\mathcal{W}}$ is invertible. ∎

74. The canonical identifications

Let the finite-dimensional linear space $\mathcal{V}$ be given. As noted in Remark 71H,(b), the fact that $\mathrm{Ev}_\mathcal{V}$ is invertible encourages us to identify $\mathcal{V}$ with $\mathcal{V}^{**}$, so that we write v instead of $\mathrm{Ev}_\mathcal{V} v$ for all $v \in \mathcal{V}$.

We note that, under the identification of $\mathbb{F}^*$ with $\mathbb{F}$ according to Remark 61B, we have

$$(\mathrm{Ev}_\mathbb{F} s)t = (\mathrm{Ev}_\mathbb{F} s)(t\otimes) = t \otimes s = st = ts = s \otimes t \quad \text{for all } s, t \in \mathbb{F};$$

this shows that the earlier identification and the one proposed here (of $\mathbb{F}$ and $\mathbb{F}^{**}$) do not clash.

Returning to an arbitrary given finite-dimensional linear space $\mathcal{V}$, we find that, under the proposed identification, (63.1) becomes

$$(74.1) \qquad v\lambda = \lambda v \quad \text{for all } v \in \mathcal{V} \text{ and } \lambda \in \mathcal{V}^*$$

(this is the reason for *not* writing v instead of $v\otimes$ for every $v \in \mathcal{V}$). Formula (61.2) applied to $\mathcal{V}^*$ instead of $\mathcal{V}$, and formula (63.2) becomes

$$(74.2) \qquad v^\mathsf{T} = v \otimes \quad \text{and} \quad (v\otimes)^\mathsf{T} = v \quad \text{for all } v \in \mathcal{V}.$$

Proposition 72C yields

$$(74.3) \qquad b^{**} = b \quad \text{for every basis } b \text{ of } \mathcal{V}.$$

If we carry out this identification for each finite-dimensional linear space, we obtain some induced identifications. In particular, if $\mathcal{V}$ and $\mathcal{W}$ are finite-dimensional linear spaces, $\mathrm{Lin}(\mathcal{V}, \mathcal{W})$ is identified with $\mathrm{Lin}(\mathcal{V}^{**}, \mathcal{W}^{**})$; Proposition 63C shows that, under this induced identification,

$$(74.4) \qquad L^{\mathsf{TT}} = L \quad \text{for all } L \in \mathrm{Lin}(\mathcal{V}, \mathcal{W}).$$

Thus each of the mappings $(L \mapsto L^\mathsf{T}) : \mathrm{Lin}(\mathcal{V}, \mathcal{W}) \to \mathrm{Lin}(\mathcal{W}^*, \mathcal{V}^*)$ and $(M \mapsto M^\mathsf{T}) : \mathrm{Lin}(\mathcal{W}^*, \mathcal{V}^*) \to \mathrm{Lin}(\mathcal{V}, \mathcal{W})$ is the inverse of the other. From (74.2) we obtain

$$(74.5) \qquad (w \otimes \lambda)^\mathsf{T} = \lambda^\mathsf{T}(w\otimes)^\mathsf{T} = \lambda \otimes w \quad \text{for all } w \in \mathcal{W} \text{ and } \lambda \in \mathcal{V}^*.$$

Again for a given finite-dimensional linear space $\mathcal{V}$ there is an induced identification of $\mathfrak{P}(\mathcal{V})$ with $\mathfrak{P}(\mathcal{V}^{**})$ by means of the bijection $\mathrm{Ev}_\mathcal{V}{}^>$ (its inverse is $\mathrm{Ev}_\mathcal{V}{}^<$). From Proposition 63G and Corollary 63H it follows that, under this induced identification,

(74.6) $$\mathcal{B}_\perp = \mathcal{B}^\perp \quad \text{for all subsets } \mathcal{B} \text{ of } \mathcal{V}^*$$

(74.7) $$\mathcal{A}^{\perp\perp} = \text{Lsp}\mathcal{A} \quad \text{for all subsets} \mathcal{A} \text{ of } \mathcal{V}.$$

The notation $\mathcal{B}_\perp$ and the concept of pre-annihilator thus become redundant under this identification.

INDEXES

Index of Terms

Index of symbols

Symbols standing for generic sets, mappings, relations, numbers, etc., are omitted when consistent with intelligibility, and do not affect alphabetic order when present.

Printed in the United States
By Bookmasters